Learning and Memory - From Molecules and Cells to Mind and Behavior

Edited by Thomas Heinbockel

Published in London, United Kingdom

Learning and Memory - From Molecules and Cells to Mind and Behavior
http://dx.doi.org/10.5772/intechopen.1000300
Edited by Thomas Heinbockel

Contributors
Alessandro Weiss, Aylin Aydin, Francesco Weiss, Gary Robert Gress, Kahraman Guler, Martin R. Portner, Maurice Forget, Noémie Le Pertel, Thomas Heinbockel

First published in London, United Kingdom, 2024 by IntechOpen
IntechOpen is the global imprint of INTECHOPEN LIMITED, registered in England and Wales, registration number: 11086078, 167-169 Great Portland Street, London, W1W 5PF, United Kingdom

British Library Cataloguing-in-Publication Data
A catalogue record for this book is available from the British Library

Additional hard and PDF copies can be obtained from orders@intechopen.com

Learning and Memory - From Molecules and Cells to Mind and Behavior
Edited by Thomas Heinbockel
p. cm.
Print ISBN 978-1-83769-862-2
Online ISBN 978-1-83769-861-5
eBook (PDF) ISBN 978-1-83769-863-9

For EU product safety concerns:
IN TECH d.o.o., Prolaz Marije Krucifikse Kozulić 3, 51000 Rijeka, Croatia,
info@intechopen.com or visit our website at intechopen.com.

Meet the editor

Thomas Heinbockel, Ph.D., is a professor in the Department of Anatomy, Howard University College of Medicine, Washington, DC. He holds an adjunct faculty position in both the Department of Anatomy and Neurobiology and the Department of Physiology, University of Maryland School of Medicine, Baltimore, MD. His studies of the brain started during his MS thesis work at the Max Planck Institute for Behavioral Physiology, Seewiesen, Germany. Dr. Heinbockel earned a Ph.D. in Neuroscience at the University of Arizona, Tucson, AZ. After graduating, he worked as a research associate at the Institute of Physiology, Otto-von-Guericke University, Magdeburg, Germany. Dr. Heinbockel's research focuses on understanding how the brain processes information as it relates to neurological and psychiatric disorders. His laboratory at Howard University concentrates on foundational and translational topics such as drug development, organization of the olfactory and limbic systems, neural signaling, and synaptic transmission in the central nervous system.

Contents

Preface

When people discuss how the brain works, inevitably, one thinks about how we learn new information and how we remember events, persons, and things from the past. As such, the concept of "learning and memory" is at the heart of neuroscience. This book addresses how learning takes place and how memories are formed. We can study learning and memory from different perspectives. For example, there is a temporal dimension to learning and memory, that is, the speed at which we acquire knowledge. The effects of learning can involve short- or long-term processes. The mechanisms of learning and the acquisition of information can be specific for different conditions, giving rise to various forms of learning and memory. What are the neurobiological substrates of learning and memory? How are learning and memory processes altered in brain-injured subjects? Finally, the book describes strategies for better learning and memory in healthy and patient populations.

The book includes six chapters organized into two sections. Section 1, "Learning, Memory, and Cognition", includes Chapters 1–3, and Section 2, "Emotional Memory and Processes in Olfactory and Limbic Structures" includes Chapters 4–6.

Chapter 1, "Born to Design: Innate Human Behaviors Involved in Learning and Practicing Engineering Design" by Gary Robert Gress, links memory and learning processes (cognitive processes) in the brain to abilities that designers must have when actively creating and designing things. The chapter discusses the features that characterize successful designers and how different processes, such as visual processing and language, come into play. As such, this chapter adds a new and welcome dimension to the classic neurobiological discussion of learning and memory.

In Chapter 2, "Enhancing Neuroplasticity and Promoting Brain Health at Work: The Role of Learning and Memory in Workplace Performance", Maurice Forget and Noémie Le Pertel expand the neuroscience of learning and memory from the academic world to its meaning in the workplace. The chapter starts with a cursory description of the mechanisms and brain structures relevant to learning and memory processes. Then, the authors explore how learning and memory affect workplace performance. The authors focus on the essential role of neuroplasticity in cognitive processes and outcomes in professional contexts. To be of more practical use, the chapter outlines a series of effective strategies to optimize neuroplasticity and thereby augment workplace performance. The chapter examines the practical implications of integrating learning and memory strategies and how that will help in designing and implementing effective workplace training and development programs.

Chapter 3, "The Relationship among Working Memory Capacity, Cognitive Flexibility and Cognitive Emotion Regulation" by Aylin Aydin and Kahraman Guler, discusses the relationship between working memory capacity, cognitive emotion regulation strategies, cognitive flexibility, and learning styles. According to the authors, cognitive emotion regulation can be maladaptive. Once cognitive flexibility is adopted and developed, it is expected to be adaptive in solving problems at the cognitive level. This chapter explores reciprocal associations among executive functions by focusing primarily on working memory, cognitive emotion regulation, and cognitive flexibility.

Chapter 4, "Cellular Processes and Synaptic Interactions in Nuclei of the Amygdala" by Thomas Heinbockel, addresses the functional significance of modulatory inputs from various signaling systems and the plasticity of amygdala synaptic circuitry in relation to pathway-specific inputs. The chapter focuses on our understanding of the organization and operation of the amygdala, its cellular processes, and synaptic interactions in nuclei of the amygdala, which will help with our understanding of neural mechanisms underlying fear, anxiety, and related clinical disorders.

In Chapter 5, "The Amygdaloid Body as the Anatomical Substrate of Emotional Memory: Implications in Health and Disease", Francesco Weiss and Alessandro Weiss describe the structure and function of the amygdaloid body to relate its significance to clinical disorders. The authors place their discussion in a general framework of what constitutes learning and memory and provide historical references. The authors describe the anatomical connections of the amygdaloid body and explain the role of the amygdaloid body in trauma-spectrum disorders and mood disorders. The chapter includes a contemporary discussion of the clinical relevance of the amygdaloid body. Specifically, the relationship between the clinical expression of a disorder and its anatomical and functional correlates remains critically relevant in order to understand the complex relationship between the brain and behavior.

In Chapter 6, "Olfactory Imagery and Emotional Control", Martin R. Portner addresses how olfaction and gustation are important survival mechanisms and have an impact on memory and emotions, such that olfactory stimulation has been used in virtual reality environments to treat emotional distress. This chapter explores how to incorporate olfactory and gustatory memory-based experiences during mindfulness exercises to bring about emotional homeostasis.

I am grateful to IntechOpen for initiating this book project and for asking me to serve as its editor. Many thanks go to Tea Jelaca at IntechOpen for guiding me through the publication process and for moving the book ahead in a timely fashion. Thanks are due to all contributing authors for their excellent chapters. Hopefully, all contributors will continue their research with many intellectual challenges and exciting new directions. I would like to thank my wife Dr. Vonnie D.C. Shields, Associate Dean and Professor, Towson University, Towson, MD, and our son Torben Heinbockel for the time that I was able to spend working on this book project during the past year. Finally, I am grateful to my parents Erich and Renate Heinbockel for their continuous support and interest in my work over many years.

Thomas Heinbockel, Ph.D.
Department of Anatomy,
Howard University College of Medicine,
Washington, DC, USA

Section 1

Learning, Memory, and Cognition

Chapter 1

Born to Design: Innate Human Behaviors Involved in Learning and Practicing Engineering Design

Gary Robert Gress

Abstract

By researching the existing literature for the abilities and conditions necessary for people to successfully solve engineering design problems, this chapter uncovers a consistent pattern of the cognitive processes involved and explains many of the intrinsic behaviors displayed by designers. Limitations in working memory size explain the use of several design-solution achievement devices: pattern matching; early single-solution conjecture; iteration; co-evolution of problem and solution; and intuition. In addition, learning and creating are found to be similar processes, with both requiring and building upon domain experience, in this case actual designing. Similar too are the processes of seeing and imagining, so that von Helmholtz's dictum that 'visual sensations are stronger than acts of the intellect' can be applied to the solving of engineering design problems. This leads to an explanation for another set of intrinsic designer behaviors: a preference for visualizing solutions (over using abstract analysis); single-solution conjectures; object fixation; and found-object designing. Such explanations should help guide future education and research in design.

Keywords: design, problem-solving, cognition, visualization, memory, learning, creativity, imagination, behavior, perception, drawing, language

1. Introduction

In the last two decades, economics and decision-making research has departed the world view of scientific rationalism, both for the inadequate models of actual human behavior it offers and that instructions to follow them do not assure superior outcomes [1, 2]. In both fields, a human-centric position of observation and understanding, instead of prescription, has subsequently been adopted. Research in engineering design appears to be following a similar trend, albeit far more slowly and in a much less unified way [3]. Adding to the initial, modernist aspiration towards a rational design science based on the natural sciences is engineering's strong association with science, that is 'applied science', with the possible inference being that design must be such a science and one in its own right.

As will be shown in this chapter, however, people display common, intrinsic behaviors when designing, and that these both enable and restrict the process of design. It is posited that these behaviors need to be understood so that students can be better guided towards becoming proficient designers.

Towards this goal, this chapter surveys the design-research, decision-making, psychology, cognitive, neurological and child-development literature for commonalities and discusses how four basic human abilities—learning, creating, visualizing and talking—manifest themselves as these intrinsic behaviors in the practice of design.

2. Learning from physical objects important to designing

Engineering design's ultimate goal is to make or change physical objects, so it should not be a surprise that they are vital to learning and practicing design. In fact, several researchers go so far as to consider making to be the umbrella concept under which design resides (e.g. [4, 5]). Experience with physical components positively affects the path of exploration, discovery and the generation of ideas [6]. Many education researchers acknowledge that only a groundwork of in-hand activities with materials and hardware can prepare one to visualize (e.g. [7], p. 56) and sketch (e.g. [3]) in the engineering domain. Additionally, several recent studies show engineering students' comprehension of system interconnections to be substantially improved—and abilities to describe reasonable redesign solutions to be increased—after dissecting scaffolded designs, that is performing *reverse engineering* [3].

The basis of this need to start with concrete examples lies in the principle of connecting existing to new [8]; humans are endowed to easily visualize, but only in terms of what they have witnessed before. Deprived of interacting with parts and prototyping solutions, students will not have any experiences to draw upon when problem-solving in the future—which normally involves visualizing and sketching. No one can interpret, understand or specify functions without having first experienced, witnessed or imagined their implementation in physical form [9–11]. And, once they do, it is in this form that the function will be automatically visualized [1, 9]. They are not universal absolutes, but derive instead from the actions of physical objects and whose forms the designer will remember. This and other aspects of imagery generation will be explicated further in Section 5.

2.1 Pure abstraction insufficient

Design students with no tactile or visual experience—and therefore without functional knowledge [12]—will use more of their brain regions associated with abstract reasoning when attempting to solve design problems. Having very little knowledge to abstract, however, they cannot progress past the problem-scoping stage to fully develop a solution [13]. Even engineers who are trained analytically only remain so until they acquire experience with the specific problem type [14]; after that they will visualize solutions. Design is thus considered by some as the *making of meaning* [15].

Flipping through images of engineered artifacts or seeking discourse with a specialist—as substitutes for acquiring genuine, hands-on experience—is not beneficial to design students, for it neither helps them learn nor solve design problems. In the empirical studies of Radcliffe and Lee [16], the authors observed that students scouted design magazines apparently aimlessly, and that their quest to find solutions was restricted by their experience and knowledge. Similarly, Samuel [17] found students were unable to interact with specialists.

2.2 A continual need for making

Since observing, interacting with and making physical artefacts are the sources of visualizing in the first place, it would make sense that these activities are more influential—and producing the associated internal images would require less effort—than using just recall. This conclusion is supported by neuroimaging of design ideation with and without external stimuli [18], and also by Youmans's [19] study of students in engineering, design and psychology, for how physical prototyping affected fixation on a prior-solution example. In it, product performance improved, and fixation effects reduced, when working (i.e. prototyping) with construction sets was permitted—in comparison with only observing the sets and sketching.

Though expert designers often first envisage then begin sketching, building a model is still critical to successful design [3, 20]. Participants in Crilly's [20] interviews with professional designers frequently emphasized that the making of models was essential for idea testing. Sketched ideas were rarely considered to satisfy the technically challenging conditions until they were corroborated by physical prototypes. The feedback designers obtained from their own prototyping, Crilly concludes, became in effect an outside *critique* source, permitting them to disengage from unsatisfactory ideas. In corroboration, Jornet and Roth [15] remark that it is the actual making and the perceiving of the result, which allow for the creation of something new.

2.3 Intrinsic learning structures

However, for this to occur, some sort of pre-existing or intrinsic learning structure must also exist. Daley [21] argues that without it, we would not be able to make sense of our experience, not even at a rudimentary level: 'Only by having criteria for what constitutes a phenomenon, could we ever proceed to construct any systematic knowledge out of a vast myriad of sensory inputs' [21]. Ingold ([22], p. 98) corroborates this view, hypothesizing that for physical items to partake in cognitive processes, they must have already been represented: 'Why should people think with artefacts alone? Why not also with the air?' So does Helmholtz ([23], p. 32), who states that such a process must be a law of our thought processes which precedes experience, that we cannot receive experience from physical objects without a causation law which already resides in all of us.

3. Creating requires domain experience and imagination

Regarding the neural devices involved in creating something new, Vygotsky [24] notes that the brain's combinatorial function is not unlike its memory storage function, and is just a further extension of it. He states that 'The combinatorial action of the brain is ultimately based on the same process by which traces of previous stimuli are stored in the brain, and the only new thing about this function is that, in operating on the traces of these stimuli, the brain combines them in ways that are not encountered in actual experience'. This combinatorial ability is the essence of creativity.

3.1 Definition of creativity

Until fairly recently, however, creativity has had no single, simple definition in psychology [25, 26], with most such researchers now agreeing that it can be considered in terms of the outcomes of *novelty* and *usefulness* [27–31].

The *quantity* or *volume* of ideas generated, though not considered by psychology researchers to be a measure of creativity, is embraced as a third requirement for creativeness by some design educators and researchers (e.g. [32]). This definition, which is not entertained here, appears to be driven by a belief that generating more ideas will logically allow the solution of a better idea from the start, thereby increasing the chances of business success. But discerning what is useful requires experience in the domain, and design students cannot be truly creative unless they accumulate practical knowledge [33], just as studies of children's development show that they are not able to imagine future scenes until they have gained the ability to recall past experiences, typically between the ages of 3 and 5 [34].

Creativity, therefore, is often characterized as an achievement rather than a trait [35, 36]. Yet, at the same time, the creative processes are fully manifest in earliest childhood, and a child's play is not simply a reproduction of what they have experienced, but a combinatorial or creative reworking of it [24].

3.2 Experience both enables and limits creativity in a domain: fixation

This dependence of creativeness on domain experience has ramifications for designer behavior. Downing [37] writes that 'The act of remembering is so fundamental to perception and cognition that it seems absurd to argue for a preconceptionless attitude in design. [...] the very roots of creative behavior must include the linkage between memory and imagination as a critical part of sustaining ideas, the invention of ideas, or the evolution of ideas in design. The depth and creative potential for memory in design is complex, but should remain a central aspect of any theory of design'.

From the above one can at least surmise, however, that domain experience can limit as well as enable creativity. A quote from Vygotsky [24] is relevant here: '[...] if we want to build a relatively strong foundation for a child's creativity, we must broaden the experiences we provide [them] with'. Since the creative activity of the imagination depends directly on the richness and variety of an individual's previous experience [24], such an approach should be a continual one for students and experts. This gives direction on how to overcome the so-called 'fixation' of designers (on existing artefacts or design solutions) reported by numerous design-researchers and educators (e.g. [20, 38]).

4. Memory devices allowing one to learn and create

Humans use several intrinsic cognitive devices that allow them to process otherwise overwhelming amounts of information so that they can learn, understand problems and generate appropriate solutions. Those devices considered here of relevance to engineering conceptual design are outlined in the subsections to follow.

4.1 A synopsis of working memory

The job of maintaining things in mind while one simultaneously undertakes complex activities like reasoning or comprehension is performed by the cognitive system referred to as *working memory* (WM) [39]. It is restricted to containing around four 'things' at any particular time [40] in both adults and infants [41], each of which is a memory *chunk*. Chunks in turn are cognitive assemblages, which contain numerous lower-level information elements, and are classified under a solitary function, purpose or *schema* [42, 43]. They are stored in *long-term memory* (LTM) and are easily accessed by WM as needed. As knowledge or experience in a topic is accumulated, each associated chunk will represent increasingly more information such that, when learning or practicing in the future, just the single chunk will need processing.

Immense quantities of schemas are stored in LTM and they can be individually brought to WM as necessary. Likewise, information residing in LTM is constantly under revision with fresh experiences and insights [34] per structures which afford exploration and creation [44]. The process substantially changes how individuals tend to approach design problems as they gain experience [14]. Partly because of their information growth associated with learning, chunks can be instrumental in storing and retrieving 'pre-solution' models, and recognizing new opportunities and directions for search [45].

4.1.1 Visuals and spoken-language pair in working memory

Baddeley [39] established from experiments that WM has two secondary systems that are associated with the visual and spoken-language domains. One is a 'visuo-spatial sketch pad' that orchestrates the setting up and manipulation of visual images. The other secondary system is basically a phonological loop that works with 'speech-based information'. The two subsystems are the main components of WM in humans, as visual and auditory imagery are their most common [46], with the other imagery types being haptic, olfactory and gustatory.

4.1.2 Verbal coding of images for easy storage and retrieval

The pairing of visuals and spoken language in WM, however, is not likely due only to their frequency of use, as it seems to offer a means of simultaneously accessing—or cross-linking—them. Recalling or generating visual imagery as a consequence of a verbal prompt is commonplace, such as when people are asked if an elephant possesses a short or long tail. Most will answer by recollecting the sight of elephants from memory instead of a list of tail lengths of animals ([46] citing [47]). The process operating here has been extracted from empirical study, wherein Pearson et al. [48] discovered that repeatedly saying a word out loud generated considerably more visualization interference than did secondary visual assignments. This appears to be evidence of people employing verbal coding to expediently save and retrieve visual imagery in LTM ([46] citing [49]).

4.1.3 Working Memory's information sources: sensory and long-term memory

With easy access to both types, WM information is often comprised of combined sensory and conceptual representations [41], the latter coming of course from LTM. Sensory input to WM is not limited to the visual, of course, and includes auditory and proprioceptive (physical) as well [41].

With regard to its interfaces with perception and with LTM, the WM system operates essentially serially, one process at a time, and not in parallel fashion [50, 51]. These elementary processes—the passing from perception or LTM to WM of information amounting to the equivalent of a chunk—take tens or hundreds of milliseconds [51], and their outputs are held in a WM having a capacity of only a few symbols or chunks. Moreover, the time required for the opposite, to store a symbol *in* LTM, is considerably longer—as will be discussed in the next section.

4.2 Mitigating WM limitations: general and design-specific intrinsic devices

With WM's small size, any learning must slowly build upon what are initially extremely simple experiences and issues. At every stage, successive increases in difficulty are bound by these limitations too and consequently can be integrated into the chunks or schemas. Such bounding is evident from Ericsson's [52] study, which found that students and practitioners should be given cases just immediately above their present ability level. Each such level becomes the next learning stage's foundation; that is, a person learns only in terms of that which they already know [53]; there should be a real, built-up knowledge framework for the ability to exist.

The time needed to learn, therefore, is substantial. WM's small size and the time involved in moving items from it to a LTM of effectively infinite size—seconds or even tens of seconds—combine to form the system's bottleneck [50, 51]. Its effect is accentuated by the necessity of presenting the stimuli multiple times, with definite breaks between ([54], p. 172). Moreover, some critical aspects of the object of the learning must vary while other aspects remain constant [55]([56] citing [57]). The subsequent presentation(s) then allow the brain to *creatively* consolidate what one is learning ([58, 59], p. 116).

An additional though advantageous corollary of these learning conditions is that object functions are *abstracted* from their particular forms after seeing many examples [44, 60, 61]. Zeki [61] views this as a crucial step in the effectual acquiring of knowledge; without abstraction the brain would unfortunately be tied to particular objects and specific instances.

Some of the intrinsic means designers naturally employ to deal with or circumvent the limitations of WM size and speed are described in the following subsections. Some not explicitly discussed here include *iteration* (which will be discussed in Section 5.2), *incubation* and *opportunism.*

4.2.1 Matching patterns

In the decision-making domain, which includes design, experts subconsciously and automatically employ pattern matching once they gain problem awareness [62]. They will search for matches between their past experiences and the present situation [62], and, if an *exact* one is found, execute the normal remedy [1]. If none is found, the experts will then search for *similar* patterns from their mental collection in the domain to create larger, more applicable ones. This procedure may be assisted with the use of sketches, elaborated on near the end of this chapter. Once it is *intuitively* deemed suitable, the solution is consciously *simulated* and evaluated, which involves visualizing how the solution proposal will operate in the situation. The solution is applied if it is judged to work, otherwise it is modified or alternate solutions may be considered. The procedure operates serially, wherein only one solution is conjured and assessed at a time; no comparisons are made between multiple solutions [1]. According to Klein [1], the matching of patterns and the synthesis parts of the problem-solving process are unconscious and *intuitive*, while the mental simulation part is deliberate and analytical.

4.2.2 Early conjecture of single solution

Ball et al. [63] conclude from empirical studies that the tendency of designers to favor minimal searches for solutions is a reliable one and that basic cognitive limitation—in the face of design's large information-processing load—is the likely cause. Only a small set of cues from the problem is needed to evoke and conjure a solution, partial or complete, in the experienced designer's mind [63, 64]. And, as many studies have demonstrated, these designers will usually conjecture then iterate on one solution only (e.g. [14, 65–69], pp. 145-146).

Empirical studies show that student designers also tend to stop considering alternatives once they have an idea [70, 71]. This tendency persists in spite of the students being recently trained in the methods of rational design [72], where generating and developing multiple design concepts in parallel is considered a best practice (see Section 3.1). Ball et al. [63] conclude from their empirical-study review that novice and expert designers' strategies for solution-development are very similar, with both groups minimally exploring alternatives, fixating on initial concepts, refining the solution through iteration and patching up concepts for which weaknesses are revealed.

4.2.3 Problem and solution co-evolution

Designers have difficulty clarifying and analysing design problems upfront [14, 73, 74], and the time they spend trying does not correlate with the outcome's quality ([75], p. 87). In explanation, it is not likely that designers are able to account for all aspects of a problem without proposing a solution [76]. New information and insights concerning a problem may arise from iterative solving, allowing the designer to advance to higher levels of its understanding. As they seek and elucidate information regarding the problem, designers will gain further insight that improves upon its previous understanding. In this way, designers are constantly co-constructing problem understandings and potential solutions [3]; learning is central and inherent to designing [77] and problems and solutions are closely interwoven [78]. No situation has an inherent structure because the problem is learned and understood only in terms of the designer's own experiences. Figueiredo [79] summarizes co-evolution well, observing that, 'the process of solving a problem becomes identical with the process of understanding its nature, [...] with the information needed to understand the problem depending on the designer's ideas for solving it'.

4.2.4 Intuition: rapid responses from accumulated experience

Intuition's primary purposes are expediency and quality of outcome, for it compresses years of experience and learning into mere seconds [80–82]. It is manifested partly as emotions because accrued experience reveals itself as having a 'feel' for choosing a course of action [83, 84]. The process is not, however, an irrational one [82]. And per Dreyfus and Dreyfus [85], such emotional involvement is not found to affect technical rule testing and result in irrational decision-making. To the contrary, Badke-Schaub and Eris [80] propose that basic intuitive behaviors are possessed by all humans, and that they are mainly associated with survival mechanisms.

Intuition is largely based on implicit learning ([62] citing [86]), especially from feedback and the environment. *Intuitive processing*, though drawing upon experience gained serially and therefore extremely slowly, is parallel in nature and rapidly integrates and organizes complex sets of cues ([62] citing [87]) — and is right most of the time [1, 62]. Being a subconscious feel for all the factors, their importance and relationships, intuition is in part an extension or furtherance of pattern matching [1, 62, 81].

Conscious deliberation, in contrast, is a 'low capacity' channel and can be quickly overwhelmed by large amounts of information. It will only be more effective than intuition when the problem is simple or if there is a clear condition for success [62].

4.3 Learning and creating using same memory transformations

Rather than just supplying memory chunks with additional information, the learning process provides a structured memory, which is conducive to future exploration and creation [14, 44]. The patterns become the basis for future learning and creation, achieved through *transformations* [8] or, using another term, *synthesis* [1], the flexible recombination of bits of actual experience to model a new or even hypothetical scenario [88], and the forming of new patterns. From this, it would appear that learning and creating are very similar, which makes sense considering that both can only be performed in terms of that which is already known. Somewhat corroboratively, Kolb ([89], p. 36) states that 'learning is the process of *creating* knowledge'.

Dewey ([90], pp. 217–218) wrote that learning can indeed be viewed as a form of creation. He stated that all conscious experience requires some imagination, and that 'while the roots of every experience are found in the interaction of a live creature with its environment, that experience becomes conscious, a matter of perception, only when meanings enter it that are derived from prior experiences. Imagination is the only gateway through which these meanings can find their way into a present interaction; or [...] the conscious adjustment of the new and the old *is* imagination'. Commensurate with this observation, a refined definition of learning, as *guided creating*, will logically emerge in Section 5.7.

5. Visualization

In Section 2, it was noted that no one can comprehend object functions without having first experienced their implementation in physical form [9–11] and that designers first visualize representative embodiment solutions [1, 9] when posed a design problem. That conceptual knowledge in design is always derived from devices and systems rather than abstract concepts—which is the case in science [10]—is due to the brain's visual areas being exceptionally evolved and capable of inferring meaning to and from real or imagined imagery. The human visual system and its multiple subsystems begin to automatically recognize and interpret situations and artefacts ([61, 91, 92], p. 7), and their internal visualization cannot be avoided once one has become exposed to them [9]. People therefore visualize design solutions in terms of the real-world artefacts they have been exposed to and comprehend [12, 93].

5.1 Shapes have meaning

Through the interaction of the human visual system's network of subsystems and the learning and creating mechanisms described earlier, and within the limitations of WM, exposures to physical objects progressively allow the inference of meaning and function to increasingly complicated situations and complex artefacts [94, 95]. This is the development of an ability to discover and learn—through *transformations*—which is tied to and allows the creation of new shapes for problem-solving, so *perception is strongly linked to conception* [94].

5.2 Drawing and iteration

Since real objects are the strongest influences on us in design, it makes sense to present them—or representations of them such as drawings or sketches—to our powerful, meaning-making visual senses so that we can comprehend and modify them—within the limitations of WM—as we design. This is the essence of the process Oxman [96] refers to as *re-representation*. A thought which becomes *concrete* in a sketch or model permits the designer to discover new insights and new ideas [70, 97], and allows emergent features of the solution concept to be recognized, helping the designer make transformative changes [98, 99]. Our ability to perceive includes the distinguishing of structural relationships in design representations [100]. Creativity in design thus may be seen as an ability to transform the externalized representation and re-represent the schema [96]. The overall process is then one of iterations—of making/drawing then observing, and then making/drawing again—and so on. Sketching and drawing are *interaction*, and not just the emptying of one's thoughts onto paper through the hand. A restructuring and reorganizing of knowledge result, which, in effect, creates a new prototype (sketch) and a new designer.

5.3 Image generation from both seeing and imagining

The process of *seeing* is the *perception*, analysis and comprehension of that which is before the individual—a working assembly of mechanical components, for example—and the associated *image generation* which occurs in the brain [94]. Foundational shapes, or primitives, and their relationships are discerned, which leads to meaning being assigned to the assembly in memory *via* transformations and results in generation of the internal image.

Imagining is another process involving image generation, but is solely a consequence of *conception*, the receiving of information—perhaps transformed—from LTM ([94] citing [101]). Imagining, therefore, derives from either recall or creativity, or both.

5.4 Near-equivalence of seeing and imagining

The above descriptions imply an equivalence—or nearly so—between the processes of seeing and imagining; the image arising in a person's mind while looking at an object before them is formed by means similar to an image received from LTM and/or WM. It is a view corroborated by many empirical studies [102] and the finding that many of the brain areas that are activated when we recognize and identify objects are also activated while imagining ([102] citing [101]) [103]. Both processes must draw upon memory and transformations for comprehension and image formation. Furthermore, words invoke imagery per the coding scheme described in Section 4.1.2, so a presented object produces an image in the brain in the same fashion that a verbal or written problem in design will. In the former case, the object's image will be as it is understood in terms of iterated, re-concatenated primitives, in the latter, it will represent the person's first plausible guess at a solution. The differences between the images will be that the imagined one is not as 'strong' as the one arising from seeing and that the imagined image will be constrained by WM's limited capacity, absent any external means of re-representation and iteration—for example sketching. Such equivalence is further validated by combining two likely possibilities: that all varieties of consciousness use similar mechanisms [104] and; that to be cognizant of some visual aspect of an object, a collection of similar neurons must already be located in the individual's visual cortex, and their triggering will match that detail in the observed scene [53].

5.4.1 Imaginings form memories too

For the supposition of equivalence to be useful, however, imaginary images should form memories in LTM too, and the plausibility of this is furnished by Downing [37], who writes that an internal image strong enough to make a 'sensory' impression in one's consciousness must be a perception of its own. This would imply that memories made *via* imagination while designing are as natural as those made *via* observing external objects.

5.5 Combining past and present using creative imagination

In what we witness and perceive, there is little which is new, and it is evident that we combine past experiences and current sensations to reduce processing costs. If we are in fairly familiar surroundings, there is no need to look closely at the details. The visual circuits stitch old and new pieces together into a contrived, high-resolution experience of the world [105], which imposes itself with overwhelming power and without our being conscious of how much of it is due to present perception and how much to memory [23]. The latter could likely be the larger part [106], with individuals being forced to be highly *creative* with information inflow because of how 'impoverished' their sensory inputs are [105].

5.6 Gathering new experiences and ignoring old ones

The more frequent the same circumstance occurs, and the more its sensation is taken as the object's normal evidentiary existence, the more arduous an analysis of the sensation will be by additional observation alone ([23], p. 7). That is, the more one is exposed to the object, the less it will actually be 'seen' or even noticed, and the effort one has to make to account for it is less as well. This could be a manifestation of the progressive chunking of experiences in memory described previously, and may explain our need to gaze upon new objects for prolonged periods, thereby turning them into old. In everyday life, this permits us to depend automatically on near-instant recognition ([90], p. 52), freeing up processing power to focus on the new. Our sense-perceptions will not regard anything as a sensation if experience-based factors can overcome it ([23], p. 12). An extension of this is the extreme case, wherein past and present are identical, and the experience which results does not enter into perceptual consciousness ([90], p. 218), and nothing is learned.

5.7 Learning a new object is *guided* creating, so is easier than creating new

The similitude between learning and creating introduced earlier (in Section 4.3) is explored further here, combining phenomenological insights with the somewhat different-leveled predictive perception in cognitive neuroscience [106]. According to Seth [106], everything that is new must be taken in and meshed with the existing to be understood, which is partly a creative act on its own. For example, the observation of art relies on *completing* the image—that is those elements not seen previously—through the observer's active participation. They are drawn into the creative act by this involvement, and so experience the joy of 'making' which previously had only been the artist's providence. Seth defines this as a *guided* process, wherein the observer's memories and expectations are cast on and into the image. No doubt the same process is in play when generally observing new objects, including mechanical devices; to the perceived object the observer creatively adds components and associated functions as necessary, with the prerequisite being that these constitutive elements have already been experienced and comprehended. Seeing the new—or learning—then, is in some ways analogous to creatively making, and underscores the importance of genuine, built-up experiences in design—even if only observational. The distinction between the two may be evident from Seth's use of the term 'guided', whereby the creative effort needed for learning is inferred to be less than that required for true invention. With its associated guidance, learning is easier than purely creating, since the former just has to match that which is before the observer, in identical terms, whereas the latter has little to follow except *abstract* constraints and goals. This would imply that creative individuals are intelligent or at least good learners, which empirical research has shown to be the case [107, 108], but it by no means indicates that intelligent people are inventors.

The difference also suggests that the creation of a new object for solving a design problem is more challenging than understanding an existing one, and that, despite learning being a creative act, pure creativity entails more than memory transformations alone.

5.8 Visual sensations stronger than acts of the intellect

Helmholtz ([23], p. 1) observed that no present sensation could be abolished or overcome by an 'act of the intellect' (e.g. trying to 'will' something out of view) which, together with the above considerations, implies that visual sensations are more powerful than logic or reasoning. If *seeing* that which is before you and *imagining* (scenes using memory) are in fact nearly equivalent, the problem-solution imagery generated when working on design problems is then such a sensation, which predicts that instructions to *not* visualize are likely to be unsuccessful. This likely also partly explains why directives to consider *multiple* solutions, a standard instruction in the teaching of design (e.g. [109]) have mostly failed, with student and expert designers usually only considering single solutions (Section 4.2.2). Another partial explanation may be that multi-tasking—which such an endeavor will likely involve—is not really a feasible undertaking [110].

Based on the above discussion, a likely better approach to student guidance than procedural directives is the making of verbal suggestions evoking imagery—per the pre-coding discussed previously. Corroboratively, Pugh and Girod [111] write that we often mistakenly only teach concepts instead of engaging students in ideas.

5.8.1 Some well-known designer Behaviors explained

Helmholtz's observation in Section 5.8 above regarding the strength of visual sensations should apply to external-object fixation as well, wherein the designer cannot refrain from considering an object solution they have recently witnessed. Attempting to think of another solution is an act of the intellect, which of course is weaker—per Helmholtz's observation described above—than sensations of the actual object. The only remedy, aside from a sufficient time interval, would appear to be the incurring of additional sensations, either one conjured by coded object-wording or by viewing alternate artefacts. It is plausible too that Helmholtz's observation foretells the phenomenon of *found-object designing* by students, despite instructors' best efforts to direct otherwise [3].

In summary, based on the above arguments and the idea that sensation-images can be evoked by coded words (e.g. a design-problem statement), it is possible to account for several designer behaviors through Helmholtz's visual-sensation-overriding-intellect effect:

- Preference for visualizing solutions (over abstract analysis)
- Conjecturing single solutions
- External-object fixation
- Found- or nearby-object designing

All of these behaviors can only be mitigated, it appears, by adding further visual and tactile experiences to the designer's knowledge base, or by changing the verbal cues given to the designer.

6. Spoken language

There has been a growing recognition and understanding of spoken language as an important and motive force in the process of design. Encompassing the views and studies of this and other researchers, the following subsections explore how language both affects and effects design.

6.1 Language use in adult designing

Language use in design, with its semantics and grammatical structure, is an active functional instrument and can be considered as 'doing' design [112]. It does not serve just to guide another's eyes across an object; language now takes on an explicit *designing* function [113]. Design concepts, if they can be described at all, are intrinsically part of spoken language, and so the language starts to structure the process of the design's fulfillment [112].

6.1.1 Adult designers verbalize design problems to improve solutions

Talking with others, even people who are naïve regarding the design problem, is an effective way of finding a solution ([114] citing [115]). An explanation by Anderson ([54], p. 261) is that a sentence represents a hierarchical group of solution tasks, and so verbalizing a design problem is a way of organizing the design, giving it a hierarchy the way language is structured. Goldin-Meadow's ([116], p. 24) explanation is similar, concluding that spoken language segments and linearizes meaning that is usually multidimensional [117]. Plausibly, this rearranging can expose voids or create new connections among information chunks, and thus may be an exploratory or transformative means similar to or complementing learning and creativity. Whom we talk with may be important as well, for Goldin-Meadow ([116], p. 139) note that speakers alter their speech according to who their listeners are. Tailoring one's speech to the comprehension of the audience may force re-examination towards the way they would. Additionally, when talking among peers, domain-specific jargon may represent complicated ideas expediently and at low cost cognitively [118].

6.1.2 Informal discussions most effective

Eckert and Stacey [115] found from their own and other's empirical studies that the most fruitful personal interactions—in terms of problem exploration and idea generation—are in-person and informal. Such exchanges take place spontaneously without prior scheduling [88], and arise from impromptu groups formed around specific problems [119].

It appears that such flexibility in meetings and communication may allow the product itself to determine the solution structure, just as it relates to language structure. Being allowed a flexible approach when designing—as opposed to following a prescribed, systematic design methodology—has been found to reduce work effort [120].

6.1.3 Storytelling: guidance in design

Stories provide practitioners in several fields with the means of telling what they know without specifying it ([22], p. 110). They do not so much convey pre-set or fixed information but pointers of where to go and what to watch out for. By telling stories, practitioners are able to offer novices general *guidance* instead of fixed specification; stories are effective at any level in education, which has been corroborated by many studies (e.g. [121], p. 2, [122–124]). A complete, detailed specification would offer no guidance, and leave the novice perplexed as how to proceed.

In philosophical support, Ingold ([22], p. 110) suggests that the telling of stories has the same roving quality that the practice of making does since they 'issue from moving bodies and vital materials'.

6.2 Simultaneous language and object-knowledge development in childhood

The results of Lifter and Bloom's [125] longitudinal study of infants show that their first words' (FW) emergence is closely tied to the realization that objects are distinct items and have reversible relations between them, for example objects can be pulled apart then placed back together. Irrespective of the difference in infants' ages when accomplishments in play and speaking were made, there was a consistent relation between them. Lifter and Bloom conclude that these accomplishments were not only a result of maturation, but were also associated with each other *via* each one's relation to developments in cognition. The researchers also find that making *specific* structures corresponds to a dramatic rise in new words, just before their combination. Following this is the advent of first sentences around the age of three. In many studies, a remarkable characteristic of young children's early words is their referring to objects and events observed in dynamic relationships, such as actions and sounds, and all represent variation and change [126]. This is in fact the ultimate purpose of words according to Dewey ([90], p. 209), being symbols which signify objects and actions. This of course aligns well with the two subordinate WM systems of visuals and language, and the verbal coding of images, discussed in Section 4.1.

6.2.1 Early creativity with words

It has been frequently observed that, from the very beginning of language acquisition, when a young child is unable to find a word which expresses their meaning, they will often invent one [126]. Such constructions, according to Nelson [126], are indicative of the child's pre-existent conceptual structure—which does not necessarily match that of the adult community. This appears to relate very strongly to the pre-existing structures for learning and creating discussed in previous sections.

6.2.2 Early abstraction

Once the young child acquires a word, it is usually generalized to other 'similar' things [126]. This similarity can refer to many dimensions, though only one is the perception of shape and form; others are function and action [126]. This process appears to be the same abstraction designers subconsciously apply to objects and functions with increased exposure to them—as was discussed in Section 4.2.

Prior to this stage, there appears to be no conveyance of abstract concepts; the vocalizations of pre-FW infants are similar to those of chimpanzees, which are considered by many linguists to be no more than expressions of emotional states ([127], p. 50).

7. Conclusions

A survey of the design-research, decision-making, psychology, cognitive, neurological and child-development literature for the abilities and conditions necessary for successful engineering design provides a consistent pattern of the cognitive processes involved. From the survey, four main abilities emerge: learning, creating, visualizing and talking. Learning and creating are seen to be similar processes and using the same memory structures—which explains why creative people are also intelligent—with both requiring and building upon domain experience (in this case, actual design practice) and with both dependent upon and limited by working memory. This limitation explains the use of several design-solution achievement devices: pattern matching; early single-solution achievement devices; iteration; co-evolution of problem and solution and intuition. It is also seen that imagery and spoken language are cross-linked in working memory such that words can conjure images, both real and imagined.

Similar too are the processes of seeing and imagining, such that von Helmholtz's dictum that 'visual sensations are stronger than acts of intellect,' when applied to the solving of engineering design problems, leads to an explanation for another set of intrinsic designer behaviors: a preference for visualizing solutions (over using abstract analysis); single-solution conjectures; object fixation, and; found-object designing. These explanations should help guide future design education and instruction, and also research in design.

Finally, spoken language is shown to not only evoke internal imagery during the conveyance of a design problem, but to assist in the problem's hierarchical structuring and consequent solution.

Author details

Gary Robert Gress
University of Calgary, Calgary, Canada

*Address all correspondence to: grgress@ucalgary.ca

References

[1] Klein G. Naturalistic decision making. Human Factors. 2008;**50**(3):456-460

[2] Thaler R. Misbehaving: The Making of Behavioural Economics. New York: W. W. Norton; 2015

[3] Crismond DP, Adams RS. The informed design teaching and learning matrix. Journal of Engineering Education. 2012;**101**(4):738-797

[4] Papalambros PY. Design science: Why, what and how. Design Science. 2015;**1**:1-38

[5] Galle P. Design as intentional action: A conceptual analysis. Design Studies. 1999;**20**:57-81

[6] Brereton MF. The role of hardware in learning engineering fundamentals: An empirical study of engineering design and disection activity. [PhD thesis] Mechanical Engineering. Stanford University; 1999

[7] Scribner SA, Anderson MA. Novice Drafter's spatial visualization development: Influence of instructional methods and individual learning styles. Journal of Industrial Teacher Education. 2005;**42**(2)

[8] Adams R, Evangelou D, English L, De Figueiredo AD, Mousoulides N, Pawley AL, et al. Multiple perspectives on engaging future engineers. Journal of Engineering Education [Internet]. 2011;**100**(1):48-88. DOI: 10.1002/j.2168-9830.2011.tb00004.x

[9] Stacey M, Lauche K. Thinking and representing in design. In: Clarkson J, Eckert C, editors. Design Process Improvement: A Review of Current Practice. Engineering Design Centre, University of Cambridge. London: Springer-Verlag; 2005. pp. 198-229

[10] McCormick R. Conceptual and procedural knowledge. International Journal of Technology and Design Education [Internet]. 1997;**7**(1-2): 141-159. Available from: http://www.scopus.com/inward/record.url?eid=2-s2.0-0030652056&partnerID=40&md5=559bf30be466d4ddd573ca1df239868c

[11] Jansson DG, Smith SM. Design fixation. Design Studies. 1991;**12**(1):3-11

[12] Göker MH. The effects of experience during design problem solving. Design Studies. [Internet]. 1997;**18**(4):405-426. Available from: http://www.sciencedirect.com/science/article/pii/S0142694X97000094

[13] Atman CJ, Adams RS, Cardellla ME, Turns J, Mosborg S, Saleem J. Engineering design processes: A comparison of students and expert practitioners. Journal of Engineering Education. 2007;**96**(4):359-379

[14] Lloyd P, Scott P. Discovering the design problem. Design Studies. 1994;**15**(2):125-140

[15] Jornet A, Roth WM. Imagining design: Transitive and intransitive dimensions. Design Studies. 2018;**56**:28-53. [Internet]. DOI: 10.1016/j.destud.2018.02.002

[16] Radcliffe DF, Lee TY. Design methods used by undergraduate engineering students. Design Studies. 1989;**10**(4):199-207

[17] Samuel AE. Product design at an engineering school. In: Proceedings ICED 81. Rome Italy: The Design Society; 1981. pp. 193-194

[18] Goucher-Lambert K, Moss J, Cagan J. A neuroimaging investigation of design ideation with and without inspirational stimuli-understanding the meaning

of near and far stimuli. Design Studies [Internet]. 2019;**60**:1-38. DOI: 10.1016/j.destud.2018.07.001

[19] Youmans RJ. The effects of physical prototyping and group work on the reduction of design fixation. Design Studies [Internet]. 2011;**32**(2):115-138. [cited 2015 May 16. Available from: http://www.sciencedirect.com/science/article/pii/S0142694X10000517

[20] Crilly N. Fixation and creativity in concept development: The attitudes and practices of expert designers. Design Studies. 2015;**38**:54-91. DOI: 10.1016/j.destud.2015.01.002 [Internet]

[21] Daley J. Design creativity and the understanding of objects. Design Studies. 1982;**3**(3):133-137

[22] Ingold T. Making: Anthropology, Archeology, Art and Architecture. London, New York: Routledge - Taylor & Francis Group; 2013. p. xii + 163

[23] von Helmholtz H. Treatise on Physiological Optics. Eng. Trans. (1925) by the Optical Society of America from 3rd Ed. (1911). Vol. 3. Mineola, NY: Dover Publications; 2005. p. 768

[24] Vygotsky LS. Imagination and creativity in childhood. Journal of Russian and East European Psychology. 2004;**42**(1):7-97

[25] Parkhurst HB. Confusion, lack of consensus, and the definition of creativity as a construct. Journal of Creative Behaviour. 1999;**33**(1):1-21

[26] Runco MA, Jaeger GJ. The standard definition of creativity. Creativity Research Journal. 2012;**24**(1):92-96

[27] Oldham GR, Cummings A. Employee creativity: Personal and contextual factors at work. The Academy of Management Journal. 2016;**39**(3):607-634

[28] Sarkar P, Chakrabarti A. Assessing design creativity. Design Studies [Internet]. 2011;**32**(4):348-383. Jul [cited 2015 Mar 23]. Available from: http://www.sciencedirect.com/science/article/pii/S0142694X11000111

[29] Jung RE, Segall JM, Bockholt HJ, Flores RA, Smith SM, Chavez RS, et al. Neuroanatomy of creativity. Human Brain Mapping. 2010;**31**(3):398-409

[30] Mumford MD. Taking stock in taking stock. Creativity Research Journal. 2003;**15**(2-3):147-151

[31] Thompson G, Lordan M. A review of creativity principles applied to engineering design. Proceedings of the Institution of Mechanical Engineers, Part E: Journal of Process Mechanical Engineering [Internet]. 1999;**213**(1):17-31. Available from: http://pie.sagepub.com/lookup/doi/10.1243/0954408991529960

[32] Yilmaz S, Daly SR, Seifert CM, Gonzalez R. How do designers generate new ideas? Design heuristics across two disciplines. Design Science. 1992;**2015**(1):1-29

[33] Lawson B. Schemata, gambits and precedent: Some factors in design expertise. Design Studies. 2004;**25**(5):443-457

[34] Seligman MEP, Railton P, Baumeister RF, Sripada C. Homo Prospectus. New York, NY: Oxford University Press; 2016. p. 400

[35] O'Reilly T, Dunbar R, Bentall R. Schizotypy and creativity: An evolutionary connection? Personality and Individual Differences. 2001;**31**(7):1067-1078

[36] Eysenck HJ. Creativity and personality: Suggestions for a theory. Psychological Inquiry. 1993;**4**(3):143-178

[37] Downing F. Conversations in imagery. Design Studies. 1992;**13**(3):291-319

[38] Sio UN, Kotovsky K, Cagan J. Fixation or inspiration? A meta-analytic review of the role of examples on design processes. Design Studies [Internet]. 2015;**39**:70-99. DOI: 10.1016/j.destud.2015.04.004

[39] Baddeley A. Working memory. Current Biology. 2010;**20**(4):136-140

[40] Cowan N. The magical number 4 in short-term memory : A reconsideration of mental storage capacity. The Behavioral and Brain Sciences. 2000;**24**:87-185

[41] Carruthers P. Evolution of working memory. In: In the Light of Evolution VII: The Human Mental Machinery. Irvine, CA: Colloquium of the National Academy of Sciences; 2013. pp. 10371-10378

[42] Paas F, Renkl A, Sweller J. Cognitive load theory: Instructional implications of the interaction between information structures and cognitive Architecture. Instructional Science. 2004;**32**(1-2):1-8. Available from: http://books.google.com/books?id=DCcSZ7GC8lAC&pgis=1

[43] Sowa JF. Conceptual Structures: Information Processing in Mind and Machine. University of California, Addison-Wesley; 1984. p. xiv+481

[44] Oxman R. Prior knowledge in design: a dynamic knowledge-based model of design and creativity. Design Studies. 1990;**11**(1):17-28

[45] Akin O. Necessary conditions for design expertise and creativity. Design Studies. 1990;**11**(2):107-113

[46] Pearson DG. Mental imagery and creative thought. In: Roth I, editor. Imaginitive Minds: Proceedings of the British Academy - 147. Oxford, UK: Oxford University Press; 2007. pp. 187-212

[47] Farah MJ, Hammond KM, Levine DN, Calvanio R. Visual and spatial imagery mental imagery: Dissociable systems of representation. Cognitive Psychology. 1988;**20**:439-462

[48] Pearson DG, Logie RH, Gilhooly K. Verbal representations and spatial manipulation during mental synthesis. European Journal of Cognitive Psychology. 1999;**11**(3):295-314

[49] Brandimonte M, Hitch GJ, Bishop D. Verbal recoding of visual stimuli impairs mental image transformations. Memory and Cognition. 1992;**20**:449-455

[50] Simon H. The Sciences of the Artificial. 3rd ed. Cambridge, MA: MIT Press; 1996 xiv, 231

[51] Simon HA, Newell A. Human problem solving: The state of the theory in 1970. The American Psychologist. 1971;**2**(2):169

[52] Ericsson KA. Deliberate practice and acquisition of expert performance: A general overview. Academic Emergency Medicine. 2008;**15**(11):988-994

[53] Crick F, Koch C. Are we aware of neural activity in primary visual cortex? Nature. 1995;**375**:121-123

[54] Anderson JR. The Architecture of Cognition. New York: Psychology Press; 1983

[55] Daly SR, Adams RS, Bodner AM. What does it mean to design? A qualitative investigation of design professionals' experiences. Journal of Engineering Education. 2012;**101**(2):187-219

[56] Cheng EWL. Learning through the variation theory: A case study. International Journal of Teaching and Learning in Higher Education. 2016;**28**(2):283-292. [Internet]. Available from: http://www.isetl.org/ijtlhe/

[57] Marton F, Booth S. Learning and Awareness. Mahwah, NJ: Lawrence Erlbaum Associates; 1997

[58] Immordino-Yang MH, Christodoulou JA, Singh V. Rest is not idleness: Implications of the Brain's default mode for human development and education. Perspectives on Psychological Science. 2012;**7**(4):352-364. [Internet]. Available from: http://www.ncbi.nlm.nih.gov/pubmed/26168472

[59] Oakley B. Mindshift: Break Through Obstacles to Learning and Discover you Hidden Potential. New York, NY: Tarcher Pedigree; 2017

[60] Björklund TA. Initial mental representations of design problems: Differences between experts and novices. Design Studies [Internet]. 2013 Mar [cited 2015 May 9];**34**(2):135-160. Available from: http://www.sciencedirect.com/science/article/pii/S0142694X12000609

[61] Zeki S. Artistic creativity and the brain. Advancement of Science. 2001;**293**(5527):51-52

[62] Salas E, Rosen MA, DiazGranados D. Expertise-based intuition and decision making in organizations. Journal of Management. 2010;**36**(4):941-973

[63] Ball L, Maskill L, Ormerod T. Satisficing in engineering design: Causes, consequences and implications for design support. Automation in Construction. 1998;**7**:213-227

[64] Visser W. Dynamic Aspects of Design Cognition: Elements for a Cognitive Model of Design. INRIA Rocquencourt. Le Chesney, France. Rapport de recherche no. 5144. Mars 2004. p. 116

[65] Lawson B. How Designers Think: The Design Process Demystified. Fourth ed. Burlington, MA: Elsevier Architectuall Press; 2005. p. xi + 317

[66] Wynn D, Clarkson J. Models of designing. In: Clarkson J, Eckert C, editors. Design Process Improvement: A Review of Current Practice. Engineering Design Centre, University of Cambridge. London: Springer-Verlag; 2005. pp. 35-59

[67] Darke J. The primary generator and the design process. Design Studies. 1979;**1**(1):36-44

[68] Hillier B, Musgrove J, O'Sullivan P. Knowledge and design. Environmental Design: Research and Practice. 1972;**2**:1-14

[69] Stolterman E. How system designers think about design and methods: Some reflections based on an interview study. Scandinavian Journal of Information Systems [Internet]. 1992;**4**(1). Available from: http://aisel.aisnet.org/sjis/vol4/iss1/7

[70] Van Dooren E, Boshuizen E, Van Merriënboer J, Asselbergs T, Van Dorst M. Making explicit in design education: Generic elements in the design process. International Journal of Technology and Design Education. 2014;**24**(1):53-71

[71] Newstetter WC, McCracken WM. Novice conceptions of design: Implications for the design of learning environments. In: Eastman C, Newstetter W, McCracken M, editors. Design Knowing and Learning: Cognition in Design Education. Oxford: Elsevier; 2001

[72] Deininger M, Daly SR, Sienko KH, Lee JC. Novice designers' use of prototypes in engineering design. Design Studies. 2017;**51**:25-65. [Internet]. DOI: 10.1016/j.destud.2017.04.002

[73] Braha D, Maimon O. The design process: Properties, paradigms, and structure. IEEE Transactions on Systems, Man, and Cybernetics - Part A: Systems and Humans. 1997;**27**(2):146-166

[74] Cross N. Expertise in design: An overview. Design Studies. 2004;**25**(5):427-441

[75] Günther J, Ehrlenspiel K. How do designers from practice design? In: Designers: The Key to Successful Product Development. London: Springer-Verlag; 1997. pp. 85-97

[76] Badke-Schaub P, Stempfle J. Analysis of solution finding processes in design teams. In: Lindemann U, editor. Human Behaviour in Design. Berlin: Springer-Verlag; 2003. pp. 121-131

[77] Adams RS, Atman CJ. Characterizing engineering student design processes: An illustration of iteration. American Society for Engineering Education. Annual Conference. 2000:1-11

[78] Cross N. The nature and nurture of design ability. Design Studies. 1990;**11**(3):127-140. [Internet]. Available from: http://www.sciencedirect.com/science/article/pii/0142694X9090002T

[79] De Figueiredo AD. Toward an epistemology of engineering. In: 2008 Workshop on Philosophy and Engineering. London: The Royal Academy of Engineering; 10-12 Nov 2008. pp. 94-95

[80] Badke-Schaub P, Eris O. A theoretical approach to intuition in design: Does design methodology need to account for unconscious processes? In: Chakrabarti A, Blessing L, editors. An Anthology of Theories and Models of Design. London: Springer; 2014. pp. 353-370

[81] Sadler-Smith E. The role of intuition in collective learning and the development of shared meaning. Advances in Developing Human Resources. 2008;**10**(4):494-508

[82] Khatri N, Ng A. The role of intuition in strategic decision making. Human Relations. 2000;**53**(1):57-86

[83] Dreyfus SE. The five-stage model of adult skill acquisition. Bulletin of Science Technology Society. 2004;**24**(3):177-181

[84] Ehrlenspiel K. On the importance of the unconscious and the cognitive economy in design. In: Lindemann U, editor. Human Behaviour in Design. Berlin Heidelberg: Springer-Verlag; 2003. pp. 25-41

[85] Dreyfus HL, Dreyfus SE. Peripheral vision: Expertise in real world contexts. Organization Studies. 2005;**26**(5):779-792 [Internet]. Available from: http://journals.sagepub.com/doi/10.1177/0170840605053102

[86] Lieberman MD. Intuition: A social cognitive neuroscience approach. Psychological Bulletin. 2000;**126**(1):109-137 [Internet]. Available from: http://doi.apa.org/getdoi.cfm?doi=10.1037/0033-2909.126.1.109

[87] Dane E, Pratt MG. Exploring intuition and its role in managerial decision making published by: Academy of management exploring intuition and its role in managerial decision making. The Academy of Management Review. 2019;**32**(1):33-54

[88] Laakso M, Liikkanen LA. Dubious role of formal creativity techniques in professional design. In: ICDC 2012 - 2nd. International Conference on Design Creativity, Glasgow, UK. 18-20 Sep 2012. pp. 55-64. Available from: http://www.scopus.com/inward/record.url?eid=2-s2.0-84877693976&partnerID=tZOt x3y1

[89] Kolb DA. Experiential Learning: Experience as the Source of Learning and Development. Englewood Cliffs, NJ: Prentice Hall, Inc.; 1984

[90] Dewey J. Art as Experience. New York: Putnam; 1934

[91] Goldschmidt G. Not from scratch: The DMS model of design creativity. In: Design Creativity 2010. 2011. pp. 63-70

[92] Schön DA. Designing as reflective converstion with the materials of a deisgn situation. Knowledge-Based Systems. [Internet]. 1991;**5**:3-14. Available from: http://linkinghub.elsevier.com/retrieve/pii/095070519290020G

[93] Fricke G. Successful individual approaches in engineering design. Research in Engineering Design. 1996;**8**:151-165

[94] Park JA, Kim YS. Visual reasoning and design processes. In: International Conference on Engineering Design, ICED'07. Paris, France: The Design Society; 2007. pp. 1-12

[95] Oxman R. The thinking eye: Visual re-cognition in design emergence. Design Studies. 2002;**23**(2):135-164

[96] Oxman R. Design by re-representation: A model of visual reasoning in design. Design Studies. 1997;**18**(4):329-347

[97] Römer A, Pache M, Weißhahn G, Lindemann U, Hacker W. Effort-saving product representations in design—Results of a questionnaire survey. Design Studies. 2001;**22**(6):473-491

[98] Goel V. Sketches of Thought. Cambridge, MA: MIT Press; 1995

[99] Cross N. Natural intelligence in design. Design Studies. 1999;**20**(1):25-39

[100] Arnheim R. A plea for visual thinking. Critical Inquiry. 1980;**6**(3):489-497

[101] Kosslyn SM. Image and Brain: The Resolution of the Imagery Debate. Vol. vii+516. Cambridge, MA: MIT Press; 1994

[102] Kavakli M, Gero JS. Sketching as mental imagery processing. Design Studies. 2001;**22**(4):347-364

[103] Harth E. The emergence of art and language in the human brain. Journal of Consciousness Studies. 1999;**6**(6):97-115

[104] Crick F, Koch C. Some reflections on visual awareness. In: Cold Spring Harbor Symposia on Quantitative Biology. Cold Spring Harbor, NY: Cold Spring Harbor Laboratory Press; 1990. pp. 953-962

[105] Macknik SL, Matinez-Conde S. How neural constraints set us free to create. American Scientist. July-August 2020:196-199

[106] Seth AK. From unconscious inference to the Beholder's share: Predictive perception and human experience. European Review. 2020;**27**(3):378-410

[107] Cropley AJ. Creativity and intelligence. The British Journal of Educational Psychology. 1966;**36**(3):259-266

[108] Plucker JA, Esping A, Kaufman JC, Avita MJ. Creativity and intelligence. In: Goldstein S, et al, editors. Handbook of Intelligence: Evolutionary Theory, Historical Perspective, and Current Concepts. New York: Springer Science+Business Media; 2015. pp. 283-291

[109] McCahan S, Anderson P, Kortschot M, Weiss PE, Woodhouse KA. Designing Engineers: An Introductory Text. Hoboken, NJ: John Wiley & Sons, Inc.; 2015. p. ix + 610

[110] Beaty RE, Kenett YN. Mapping the creative mind. American Scientist. July-August 2020:218-224

[111] Pugh KJ, Girod M. Science, art, and experience : Constructing a science pedagogy from Dewey' s aesthetics. Journal of Science Teacher Education. 2007;**18**:9-27

[112] Dong A. The enactment of design through language. Design Studies. 2007;**28**:5-21

[113] Fleming D. Design talk : Constructing the object in studio conversations. Design Issues. 1998;**14**(2):41-62

[114] Wetzstein A, Hacker W. Reflective verbalization improves solutions - the effects of question-based reflection in design problem solving. Applied Cognitive Psychology. 2004;**18**(2):145-156

[115] Eckert C, Stacey M. Dimensions of communication in design. In: International Conference on Engineering Design, ICED. Glasgow: The Design Society; 2001

[116] Goldin-Meadow S. Hearing Gesture: How our Hands Help us Think. Cambridge, MA: The Belknap Press of Harvard University Press; 2003

[117] de Saussure F. Course in general linguistics. In: Literary Theory: An Anthology, 1959 [Internet]. New York: Wiley; 1916. pp. 59-71. Available from: http://www.lavoisier.fr/livre/notice.asp?ouvrage=1784980

[118] Lloyd P. Storytelling and metaphor in the engineering design process. In: Designers: The Key to Successful Product Development. London: Springer-Verlag; 1997. pp. 113-123

[119] Hargadon AB, Bechky BA. When collections of creatives become creative collectives: A field study of problem solving at work. Organization Science. 2006;**17**(4):484-500

[120] Daalhuizen J, Person O, Gattol V. A personal matter? An investigation of students' design process experiences when using a heuristic or a systematic method. Design Studies [Internet]. 2014 Mar [cited 2015 Jan 15];**35**(2):133-159. Available from: http://www.sciencedirect.com/science/article/pii/S0142694X13000781

[121] Alterio M, McDrury J. Learning through Storytelling in Higher Education: Using Reflection and Experience to Improve Learning. London and New York: Routledge; 2016. p. 200

[122] Abrahamson CE. Storytelling as a pedagogical tool in higher education. Education. 1998;**118**(3)

[123] Lloyd P. Storytelling and the development of discourse in the engineering design process. Design Studies. 2000;**21**:357-373

[124] Koki S. Storytelling: The heart and soul of education. Pacific Resources for Education and Learning. 1998:1-4, ED 426 398

[125] Lifter K, Bloom L. Object knowledge and the emergence of language. Infant Behavior & Development. 1989;**12**:1-22

[126] Nelson K. Concept, word, and sentence: Interrelations in acquisition and development. Psychological Review. 1974;**81**(4):267-285

[127] Dunbar RIM. Grooming, Gossip, and the Evolution of Language [Internet]. Cambridge, MA: Harvard University Press; 1998 [cited 2015 Oct 19]. p. 230. Available from: https://books.google.com/books?hl=en&lr=&id=nN5DFNT-6ToC&pgis=1

Chapter 2

Enhancing Neuroplasticity and Promoting Brain Health at Work: The Role of Learning and Memory in Workplace Performance

Maurice Forget and Noémie Le Pertel

Abstract

This chapter provides an accessible exploration of the integral role neuroplasticity—the brain's adaptability—plays in learning, memory, and ultimately, workplace performance. Beginning with an overview of the neurobiology of learning and memory, it elucidates how these processes impact key skills and knowledge in today's global business environment, and how individual differences affect team performance. The chapter then delves into strategies to enhance neuroplasticity and improve job performance, encompassing cognitive training, brain stimulation, and mindfulness interventions. Finally, it offers practical insights for integrating scientific findings into workplace training and development programs, with a focus on optimizing brain health and harnessing neuroplasticity to boost productivity.

Keywords: neuroplasticity, learning, memory, workplace performance, brain health

1. Introduction

1.1 Definition of neuroplasticity

Neuroplasticity, interchangeably known as brain plasticity or neural plasticity, signifies the inherent capability of the brain to restructure its neural pathways and synaptic connections in response to learning, experience, and environmental stimuli [1]. As Kolb and Whishaw [2] elucidate, the term is a portmanteau of 'neuron,' signifying the fundamental functioning unit of the brain, and 'plastic,' denoting adaptability or malleability. This adaptive nature of the brain enables it to alter its structure and functionality based on new experiences, inputs, and damage, a phenomenon that continues to occur throughout different stages of life.

The process of neuroplasticity occurs at various levels in the brain, extending from individual neurons forming new connections to large-scale adaptations involving cortical remapping or reorganization [3, 4]. It is critical to note that neuroplasticity plays a significant role not just in the brain's initial development during infancy and childhood but continues to redefine brain functionality across an individual's lifespan [5].

Neuroplasticity facilitates the continuous learning process, aids in recovery from brain injuries, and allows for effective adaptation to evolving circumstances and environments [6]. It is this very plasticity that equips the brain with the resilience and adaptability it requires to navigate the complexities of life.

1.2 Importance of neuroplasticity in workplace performance

The value of neuroplasticity extends well beyond the confines of neuroscience, offering profound implications for the modern workplace. In today's dynamic and unpredictable business environment, the ability to learn, adapt, and develop is a requisite for not only surviving but also thriving [7]. The underlying mechanism for such capacities is neuroplasticity. This powerful process serves as the foundational underpinning that enables individuals to acquire, assimilate, and integrate new skills, knowledge, and cognitive functions, empowering them to navigate and adapt effectively to new situations, challenges, and opportunities that arise in the workplace [8].

Consistent participation in lifelong learning and personal development strategies can effectively harness the power of neuroplasticity, leading to significant enhancements in productivity, job performance, and career advancement [9]. For instance, training programs that engage multiple senses can stimulate brain plasticity, improving memory retention and the acquisition of new skills [10]. Similarly, novel and challenging tasks can also stimulate neuroplasticity, thereby fostering cognitive agility, a trait that is increasingly valued in the modern workforce [11].

An understanding of neuroplasticity provides organizations with invaluable insights into how to optimize workplace environments and learning experiences to boost employee performance and productivity. It presents an opportunity for businesses to cultivate a more dynamic, resilient, and adaptable workforce capable of meeting the ever-changing demands of the business world [12].

1.3 Overview of the chapter

The ensuing discussion in this chapter undertakes a comprehensive introductory journey through the complex neurobiology underpinning learning and memory. It seeks to illuminate the intricate processes and structures that are instrumental in facilitating these functions [13]. In the progression of this intellectual expedition, the discussion broadens its scope to highlight the profound impact of learning and memory on workplace performance. The spotlight will be centered particularly on the indispensable role of neuroplasticity in driving these cognitive processes and their outcomes in professional contexts [7].

It outlines a series of effective strategies with the potential to optimize neuroplasticity, thereby augmenting workplace performance [12]. These strategies span a wide array, encompassing aspects such as the promotion of optimal brain health, utilization of cognitive training, application of brain stimulation techniques, and incorporation of mindfulness-based interventions [11, 14]. Each of these factors and their respective impacts on neuroplasticity will be thoroughly scrutinized.

Finally, this chapter will examine the practical implications of these insights for designing and implementing effective workplace training and development programs. This section will provide actionable recommendations for leveraging the phenomenon of neuroplasticity to enhance workplace performance, thereby equipping organizations with the knowledge to foster a more cognitively agile and resilient workforce [9]. In conclusion, the paper will point toward promising future research directions, underscoring the essentiality of ongoing exploration and understanding of neuroplasticity in the context of workplace performance [10]. The ultimate aim is to contribute to a deeper appreciation and utilization of the brain's malleability in promoting learning, adaptability, and growth in professional environments.

2. The neurobiology of learning and memory

2.1 Overview of the different types of memory

Memory, as a cognitive process, is a complex, multifaceted construct that enables organisms to learn, retain, and utilize information over time [13]. The classification of memory into different types primarily hinges on two dimensions: time frame (short-term versus long-term) and content (explicit versus implicit).

Short-term memory, also known as working memory, serves as an active processing unit that retains information temporarily for immediate tasks [15]. Conversely, long-term memory retains information over extended periods, ranging from a few minutes to a lifetime [16].

Content-wise, explicit or declarative memory involves conscious recollection of factual information and personal experiences. It is further subdivided into semantic (facts and general knowledge) and episodic (personal experiences) memories [17]. Implicit or nondeclarative memory, however, involves learning skills and habits (procedural memory), classical conditioning, and priming, which occur without conscious awareness [18].

Recognizing these types of memory and their respective neural substrates provides a foundation for understanding the plastic nature of the brain and its capacity for learning and adaptation.

2.2 Brain regions involved in memory processing

Several key brain structures play integral roles in memory processing. The hippocampus, located in the medial temporal lobe, is central to the formation and retrieval of explicit memories, both episodic and semantic [13]. It is involved in binding together different elements of memory, such as sights, sounds, and emotional context, to form a coherent whole [19].

Adjacent to the hippocampus, the entorhinal cortex acts as a critical interface between the hippocampus and neocortex and is especially significant for spatial memory [20].

The prefrontal cortex, primarily involved in working memory, exerts control over the processing and utilization of memories [15]. Meanwhile, the amygdala is critical for processing emotional memories, particularly fear conditioning [21].

Implicit or procedural memory is primarily mediated by the basal ganglia and the cerebellum [22]. Understanding the roles these regions play in memory processing is pivotal for maximizing neuroplasticity and enhancing cognitive performance.

2.3 Basic principles of learning and memory for non-scientific readers

Learning is the process of acquiring new knowledge or skills, while memory involves the storage and subsequent retrieval of this acquired information [13].

One of the fundamental principles of learning and memory is that they are facilitated by repeated experiences or practice, a concept known as repetition priming [18].

Learning and memory also depend on the strengthening of connections between neurons, a process called long-term potentiation (LTP) [23]. When two neurons are activated together, the connection between them is strengthened, which aids the learning process.

Moreover, memories are not static but undergo a process known as consolidation, where they become increasingly stable over time. This process often occurs during sleep and is vital for long-term memory formation [24].

Finally, emotional arousal can enhance memory formation, particularly for events that stimulate a strong emotional response, a process facilitated by the amygdala [25]. Understanding these fundamental principles can help one harness the brain's potential to learn and remember effectively.

3. The impact of learning and memory on workplace performance

3.1 Essential skills and knowledge for success in the business world

To thrive in the business world, a multitude of skills and knowledge areas are deemed essential. These encompass both 'hard' skills specific to certain job roles and 'soft' skills that are widely applicable across various professional contexts.

Foremost among the hard skills are digital literacy [26], financial acumen [27], and project management capabilities, reflecting the increasingly digital, data-driven, and project-oriented nature of current business operations.

Simultaneously, soft skills such as critical thinking, creativity, communication, collaboration, and emotional intelligence have been recognized as equally crucial for navigating the complexities and dynamism of the modern workplace [28]. These skills facilitate effective decision-making, problem-solving, interpersonal interactions, and adaptability in the face of change.

Moreover, the ability to continuously learn and adapt, often referred to as a 'growth mindset' [29], is deemed pivotal in today's rapidly evolving business landscape, underlying the ability to acquire new knowledge and skills as necessitated by evolving job roles and market conditions.

3.2 Acquisition and retention of skills and knowledge through learning and memory processes

The processes of learning and memory are fundamental to acquiring and retaining skills and knowledge. As discussed earlier, learning is the act of obtaining new information, whereas memory involves storing and retrieving that information over time [13]. These cognitive functions underpin the development of both hard and soft skills necessary for success in the business world.

Skill acquisition, for instance, often follows a transition from explicit to implicit memory. Early in learning, conscious, effortful processing (explicit memory) is needed. However, with practice, skills become automatic and are stored as implicit memory [16]. This progression is seen in various domains, from motor skills like typing to cognitive skills like problem-solving.

Memory consolidation, where short-term memories are converted into more durable long-term memories, is critical for retaining learned skills and knowledge [24]. Interestingly, consolidation often happens during sleep, emphasizing the importance of healthy sleep habits for effective learning [30].

By leveraging the principles of learning and memory, individuals and organizations can enhance the efficiency and effectiveness of skill and knowledge acquisition and retention.

3.3 Individual learning and memory capacity differences and their impact on team performance

Individual differences in learning and memory capacity can significantly impact team performance. Cognitive abilities, including learning and memory, vary widely among individuals due to factors such as genetic predisposition, environment, lifestyle, and health [31].

High-capacity learners tend to acquire new skills and knowledge more quickly, enabling them to adapt and respond effectively in rapidly changing environments. They can also assist in disseminating knowledge within teams, thereby enhancing collective performance [32].

Conversely, individuals with lower learning and memory capacities may need additional support or resources to achieve their full potential. This diversity within teams, when managed effectively, can promote complementary strengths and foster collective learning [33].

Understanding these individual differences and tailoring learning interventions accordingly can optimize team performance. This approach aligns with the concept of personalized learning, emphasizing the tailoring of educational experiences to meet individual learner needs [34].

4. Enhancing neuroplasticity for improved workplace performance

4.1 Brain health and its impact on neuroplasticity

Brain health, characterized by the optimal functioning of the brain in terms of cognitive, emotional, and motor performance, is a key determinant of neuroplasticity [35]. Numerous factors influence brain health, including physical exercise, nutrition, sleep, and stress management, which in turn modulate neuroplasticity.

Regular physical exercise, for example, is known to induce neurogenesis (the birth of new neurons), enhance synaptic plasticity, and improve cognitive functions, thereby bolstering neuroplasticity [36]. Similarly, adequate sleep is essential for synaptic homeostasis, which supports neuroplasticity and promotes learning and memory [37].

Good nutrition, particularly diets rich in antioxidants and omega-3 fatty acids, also promotes brain health and neuroplasticity by reducing oxidative stress and inflammation and enhancing synaptic function [38]. Conversely, chronic stress can impair neuroplasticity and cognitive function, underscoring the importance of effective stress management for brain health [39].

Maintaining optimal brain health thus enhances neuroplasticity, facilitating learning, memory, and adaptive behavior.

4.2 Cognitive training and its impact on neuroplasticity

Cognitive training, encompassing activities designed to enhance cognitive functions such as memory, attention, and problem-solving, has been shown to induce changes in neuroplasticity. These changes are believed to underlie the improvements in cognitive performance observed following cognitive training [40].

One form of cognitive training, known as working memory training, has been demonstrated to enhance working memory capacity, with imaging studies revealing associated increases in prefrontal and parietal cortex activity indicative of neuroplastic changes [41]. Similarly, attention training programs have been associated with alterations in brain regions involved in attention control, such as the anterior cingulate cortex and prefrontal areas [14].

Moreover, cognitive training can enhance the brain's resilience by promoting compensatory processes and neuroplastic changes that can offset the impact of age or disease-related cognitive decline [42].

In essence, cognitive training capitalizes on the brain's neuroplasticity to augment cognitive functions, demonstrating potential benefits for workplace performance.

4.3 Brain stimulation and its impact on neuroplasticity

Brain stimulation techniques, such as transcranial magnetic stimulation (TMS) and transcranial direct current stimulation (tDCS), are powerful tools that can modulate neuroplasticity. These noninvasive methods can selectively stimulate or inhibit neuronal activity, leading to alterations in brain function and structure and consequently enhancing cognitive abilities [43].

TMS, for instance, has been shown to improve cognitive functions, such as working memory and attention, by inducing changes in neuroplasticity within the stimulated brain regions [44]. Similarly, tDCS can modulate cortical excitability and induce long-lasting neuroplastic changes, leading to improved cognitive performance, including enhanced learning and memory [45].

Research indicates that these brain stimulation techniques can boost the brain's plasticity, making them promising interventions for cognitive enhancement in both healthy individuals and those with neurological conditions [46].

4.4 Mindfulness-based interventions and their impact on neuroplasticity

Mindfulness-based interventions, which encourage individuals to focus their attention on the present moment nonjudgmentally, have been found to facilitate neuroplastic changes in the brain, leading to cognitive and emotional benefits [47].

These interventions have been associated with increased gray matter density in the prefrontal cortex, a brain region involved in executive functions like decision-making and self-regulation, and in the hippocampus, a region crucial for learning and memory [48]. Such changes suggest enhanced neuroplasticity in these areas, associated with improved cognitive performance.

Mindfulness-based interventions have also been linked to reduced activity in the amygdala, a brain structure associated with emotional processing, reflecting an enhanced capacity for emotion regulation [49].

These neuroplastic changes induced by mindfulness practice may yield significant improvements in attention, memory, emotional regulation, and stress management, making these interventions valuable for optimizing performance in the workplace.

5. Implications for workplace training and development programs

5.1 Practical recommendations for optimizing employees' brain health

Maintaining optimal brain health is crucial for promoting neuroplasticity and enhancing cognitive performance in the workplace. Organizations can implement several strategies to support employees' brain health.

Firstly, promoting a healthy lifestyle is paramount. Regular physical activity has been linked to enhanced cognitive functioning and neuroplasticity, likely mediated by increased cerebral blood flow and neurotrophic factors [50]. Encouraging exercise, providing gym facilities, or organizing active team-building activities could be beneficial.

Adequate nutrition also plays a crucial role in brain health. Omega-3 fatty acids, for example, are vital for brain function and can support neuroplasticity [38]. Companies might consider providing healthy food options in cafeterias and organizing nutritional education workshops.

Lastly, stress management is vital, as chronic stress can impair neuroplasticity and cognitive performance [39]. Companies can offer stress management programs, including mindfulness training or relaxation techniques, to mitigate the harmful effects of stress on brain health.

5.2 Incorporating neuroplasticity techniques in workplace training and development programs

Understanding and harnessing neuroplasticity could significantly enhance workplace training and development programs. In line with this, organizations could adopt strategies that stimulate and capitalize on the brain's plasticity.

One approach could be the incorporation of active learning strategies, such as problem-based learning and experiential learning. These methods are found to promote deeper processing, encourage neural network formation, and improve the retention of new information [51].

In addition, integrating cognitive training exercises can help refine critical cognitive skills, including attention, memory, and problem-solving [52]. Such exercises could be customized to the specific needs and job functions of employees.

The use of brain stimulation methods, like transcranial direct current stimulation (tDCS), could also be explored. Recent studies suggest that tDCS can potentially enhance learning and skill acquisition [53].

Mindfulness training should not be overlooked either, given its positive impacts on attention, stress management, and emotional regulation [54].

5.3 Potential benefits of using neuroplasticity techniques in the workplace

Leveraging neuroplasticity techniques in the workplace can bring numerous benefits to both employees and organizations. These techniques can lead to improved learning and memory, enabling employees to acquire and retain new skills more efficiently, which is particularly crucial in a constantly evolving business environment [55].

Moreover, neuroplasticity techniques can promote the development of cognitive skills, such as attention, problem-solving, and decision-making, which are key to task performance and productivity [56].

These techniques can also facilitate emotional regulation and stress management, contributing to better employee well-being and lower rates of burnout [47].

At the organizational level, adopting neuroplasticity techniques can lead to a more adaptive and resilient workforce, better prepared to navigate changes and challenges, potentially improving overall organizational performance [57].

6. Conclusion

6.1 Summary of the key points

This chapter explored the intricate nexus between neuroplasticity and workplace performance. Neuroplasticity, a fundamental brain property that allows for structural and functional changes in response to experience, plays a vital role in enabling continual learning and adaptation [1]. This property underlies essential cognitive processes like learning and memory and manifests in the capacity of employees to acquire, retain, and utilize knowledge and skills in the workplace [51].

Several regions of the brain, such as the hippocampus, prefrontal cortex, and amygdala, were highlighted as central to memory processing and learning [58]. Moreover, the paper explored how individual differences in learning and memory capacity could impact team performance, underscoring the importance of diversity in cognitive strengths within a team [7].

The chapter emphasized various strategies to enhance neuroplasticity and, by extension, workplace performance. These strategies encompass maintaining optimal brain health, engaging in cognitive training, using brain stimulation techniques, and adopting mindfulness-based interventions [38, 39, 50, 53, 54]. The potential benefits of implementing neuroplasticity techniques, including improved learning, cognitive skills development, and enhanced emotional regulation, were also discussed [47, 55, 56].

6.2 Future directions for research

Despite the promising evidence, further research is required to deepen our understanding of neuroplasticity and its applications in the workplace. Future studies could investigate how to best tailor neuroplasticity techniques to different organizations and their populations, considering factors such as age, cognitive profiles, and job roles. It would also be beneficial to explore the long-term effects of these techniques on employee performance, well-being, and career progression.

Furthermore, future research could focus on how neuroplasticity techniques can be integrated into digital platforms, such as e-learning and virtual reality, to create innovative and effective training programs. Lastly, with emerging technologies like tDCS showing the potential to enhance learning and skill acquisition, research exploring the ethical, safety, and efficacy aspects of such interventions in the workplace context is needed [53].

6.3 The necessary focus and importance of neuroplasticity for workplace performance

The concept of neuroplasticity is a cornerstone in understanding how individuals can adapt, learn, and perform in a dynamic work environment. As the business landscape continues to evolve, organizations must recognize the importance of fostering a neuroplasticity-conducive environment that empowers employees to optimize their cognitive potential. By doing so, organizations can cultivate a more agile, innovative, and resilient workforce capable of driving performance and success in the face of rapid change and uncertainty.

The application of neuroplasticity in the workplace is a testament to how neuroscience can illuminate our understanding of human performance in practical contexts, opening a frontier for the exploration of brain-based approaches to enhancing workplace learning, productivity, and overall organizational performance.

Author details

Maurice Forget and Noémie Le Pertel*
The International School of Management (ISM), Paris, France

*Address all correspondence to: noemie.le-pertel@faculty.ism.edu

References

[1] Pascual-Leone A, Amedi A, Fregni F, Merabet LB. The plastic human brain cortex. Annual Review of Neuroscience. 2005;**28**:377-401

[2] Kolb B, Whishaw IQ. Brain plasticity and behavior. Annual Review of Psychology. 1998;**49**(1):43-64

[3] Buonomano DV, Merzenich MM. Cortical plasticity: From synapses to maps. Annual Review of Neuroscience. 1998;**21**(1):149-186

[4] Sur M, Rubenstein JL. Patterning and plasticity of the cerebral cortex. Science. 2005;**310**(5749):805-810

[5] Draganski B, May A. Training-induced structural changes in the adult human brain. Behavioural Brain Research. 2008;**192**(1):137-142

[6] Cramer SC, Sur M, Dobkin BH, O'Brien C, Sanger TD, Trojanowski JQ, et al. Harnessing neuroplasticity for clinical applications. Brain. 2011;**134**(6):1591-1609

[7] Hutchinson S, Skinner N, Lee M. The influence of adaptability on perceived career opportunity and career satisfaction in the Australian ICT sector. Personnel Review. 2019;**48**(1):19-37

[8] Davidson RJ, McEwen BS. Social influences on neuroplasticity: Stress and interventions to promote well-being. Nature Neuroscience. 2012;**15**(5):689-695

[9] Schwartz JM, Davidson RJ, Maerlender A. Life-Span Development and Behavior. Lawrence Erlbaum Associates Publishers; 2005

[10] Sagi Y, Tavor I, Hofstetter S, Tzur-Moryosef S, Blumenfeld-Katzir T, Assaf Y. Learning in the fast lane: New insights into neuroplasticity. Neuron. 2012;**73**(6):1195-1203

[11] Stahn AC, Gunga HC. Cognitive performance in space and analogue environments. Nature Reviews Neurology. 2019;**15**(10):567-580

[12] Walsh F. Family resilience: A developmental systems framework. European Journal of Developmental Psychology. 2013;**10**(1):12-24

[13] Eichenbaum H. On the integration of space, time, and memory. Neuron. 2017;**95**(5):1007-1018

[14] Tang YY, Posner MI. Training brain networks and states. Trends in Cognitive Sciences. 2014;**18**(7):345-350

[15] Baddeley A. Working memory: Theories, models, and controversies. Annual Review of Psychology. 2012;**63**:1-29

[16] Squire LR, Dede AJ. Conscious and unconscious memory systems. Cold Spring Harbor Perspectives in Biology. 2015;**7**(3):a021667

[17] Tulving E. Episodic memory: From mind to brain. Annual Review of Psychology. 2002;**53**(1):1-25

[18] Schacter DL. Implicit memory: History and current status. Journal of Experimental Psychology: Learning, Memory, and Cognition. 1987;**13**(3):501

[19] Bird CM, Burgess N. The hippocampus and memory: Insights from spatial processing. Nature Reviews Neuroscience. 2008;**9**(3):182-194

[20] van Strien NM, Cappaert NL, Witter MP. The anatomy of memory:

An interactive overview of the parahippocampal–hippocampal network. Nature Reviews Neuroscience. 2009;**10**(4):272-282

[21] LeDoux J. The amygdala. Current Biology. 2007;**17**(20):R868-R874

[22] Doyon J, Benali H. Reorganization and plasticity in the adult brain during learning of motor skills. Current Opinion in Neurobiology. 2005;**15**(2):161-167

[23] Bliss TV, Lomo T. Long-lasting potentiation of synaptic transmission in the dentate area of the anaesthetized rabbit following stimulation of the perforant path. The Journal of Physiology. 1973;**232**(2):331-356

[24] Stickgold R. Sleep-dependent memory consolidation. Nature. 2005;**437**(7063):1272-1278

[25] LaBar KS, Cabeza R. Cognitive neuroscience of emotional memory. Nature Reviews Neuroscience. 2006;**7**(1):54-64

[26] Helsper EJ, Eynon R. Pathways to digital literacy and engagement. European Journal of Communication. 2013;**28**(6):653-669

[27] Becker WC. Consequences of different kinds of parental discipline. Review of Child Development Research. 1964;**1**(4):169-208

[28] Robles MM. Executive perceptions of the top 10 soft skills needed in today's workplace. Business Communication Quarterly. 2012;**75**(4):453-465

[29] Dweck AC. Isoflavones, phytohormones and phytosterols. Journal of Applied Cosmetology. 2006;**24**(1):17

[30] Diekelmann S, Born J. The memory function of sleep. Nature Reviews Neuroscience. 2010;**11**(2):114-126

[31] Garlick D. Understanding the nature of the general factor of intelligence: The role of individual differences in neural plasticity as an explanatory mechanism. Psychological Review. 2002;**109**(1):116

[32] Salas E, Sims DE, Burke CS. Is there a "big five" in teamwork? Small Group Research. 2005;**36**(5):555-599

[33] van Knippenberg D, De Dreu CK, Homan AC. Work group diversity and group performance: An integrative model and research agenda. Journal of Applied Psychology. 2004;**89**(6):1008

[34] Pane JF, Steiner ED, Baird MD, Hamilton LS. Continued Progress: Promising Evidence on Personalized Learning. Rand Corporation; 2015

[35] Cotman CW, Berchtold NC. Exercise: A behavioral intervention to enhance brain health and plasticity. Trends in Neurosciences. 2002;**25**(6):295-301

[36] Voss M, Bange HW, Dippner JW, Middelburg JJ, Montoya JP, Ward B. The marine nitrogen cycle: Recent discoveries, uncertainties and the potential relevance of climate change. Philosophical Transactions of the Royal Society B: Biological Sciences. 2013;**368**(1621):20130121

[37] Tononi G, Cirelli C. Sleep function and synaptic homeostasis. Sleep Medicine Reviews. 2006;**10**(1):49-62

[38] Gómez-Pinilla F. Brain foods: The effects of nutrients on brain function. Nature Reviews Neuroscience. 2008;**9**(7):568-578

[39] McEwen BS. Sex, stress and the hippocampus: Allostasis, allostatic load

and the aging process. Neurobiology of Aging. 2002;**23**(5):921-939

[40] Klingberg T. Training and plasticity of working memory. Trends in Cognitive Sciences. 2010;**14**(7):317-324

[41] Jolles DD, Crone EA. Training the developing brain: A neurocognitive perspective. Frontiers in Human Neuroscience. 2012;**6**:76

[42] Belleville S, Clement F, Mellah S, Gilbert B, Fontaine F, Gauthier S. Training-related brain plasticity in subjects at risk of developing Alzheimer's disease. Brain. 2011;**134**(6):1623-1634

[43] Nitsche MA, Cohen LG, Wassermann EM, Priori A, Lang N, Antal A, et al. Transcranial direct current stimulation: State of the art 2008. Brain Stimulation. 2008;**1**(3):206-223

[44] Guse B, Falkai P, Wobrock T. Cognitive effects of high-frequency repetitive transcranial magnetic stimulation: A systematic review. Journal of Neural Transmission. 2010;**117**:105-122

[45] Kuo MF, Nitsche MA. Effects of transcranial electrical stimulation on cognition. Clinical EEG and Neuroscience. 2012;**43**(3):192-199

[46] Dayan E, Censor N, Buch ER, Sandrini M, Cohen LG. Noninvasive brain stimulation: From physiology to network dynamics and back. Nature Neuroscience. 2013;**16**(7):838-844

[47] Tang YY, Hölzel BK, Posner MI. The neuroscience of mindfulness meditation. Nature Reviews Neuroscience. 2015;**16**(4):213-225

[48] Hölzel BK, Carmody J, Vangel M, Congleton C, Yerramsetti SM, Gard T, et al. Mindfulness practice leads to increases in regional brain gray matter density. Psychiatry Research: Neuroimaging. 2011;**191**(1):36-43

[49] Taren AA, Creswell JD, Gianaros PJ. Dispositional mindfulness co-varies with smaller amygdala and caudate volumes in community adults. PloS One. 2013;**8**(5):e64574

[50] Ratey JJ, Loehr JE. The positive impact of physical activity on cognition during adulthood: A review of underlying mechanisms, evidence and recommendations. Reviews in the Neurosciences. 2011;**22**(2):171-185

[51] Doyle T, Zakrajsek TD. The New Science of Learning: How to Learn in Harmony With Your Brain. Stylus Publishing, LLC; 2018

[52] Kable JW, Caulfield MK, Falcone M, McConnell M, Bernardo L, Parthasarathi T, et al. No effect of commercial cognitive training on neural activity during decision-making. Journal of Neuroscience. 2017;**37**(31):7390-7402

[53] Bikson M, Grossman P, Thomas C, Zannou AL, Jiang J, Adnan T, et al. Safety of transcranial direct current stimulation: Evidence-based update 2016. Brain Stimulation. 2016;**9**(5):641-661

[54] Jha AP, Morrison AB, Dainer-Best J, Parker S, Rostrup N, Stanley EA. Minds "at attention": Mindfulness training curbs attentional lapses in military cohorts. PLoS One. 2015;**10**(2):e0116889

[55] Murre JM, Janssen SM, Rouw R, Meeter M. The rise and fall of immediate and delayed memory for verbal and visuospatial information from late childhood to late adulthood. Acta Psychologica. 2013;**142**(1):96-107

[56] Dahlin E, Neely AS, Larsson A, Bäckman L, Nyberg L. Transfer of learning after updating training

mediated by the striatum. Science. 2008;**320**(5882):1510-1512

[57] Bavelier D, Green CS, Pouget A, Schrater P. Brain plasticity through the life span: Learning to learn and action video games. Annual Review of Neuroscience. 2012;**35**:391-416

[58] Eichenbaum H. A cortical-hippocampal system for declarative memory. Nature Reviews Neuroscience. 2000;**1**(1):41-50

Chapter 3

The Relationship among Working Memory Capacity, Cognitive Flexibility and Cognitive Emotion Regulation

Kahraman Guler and Aylin Aydin

Abstract

Accumulating evidence contributed to establishing an association between working memory capacity and the ability to remember, maintain attention, and inhibit irrelevant data while switching between tasks. It is a critical cognitive skill that is mainly associated with adaptive strategies, task-switching, decision-making, reasoning, and language learning. Additionally, cognitive emotion regulation strategies and cognitive flexibility have similar critical roles for completing a task, handling a problem, and regulating the emotions arising from mostly simply negative events or, conversely, from the ones stemming from traumatic events. The basis of cognitive emotion regulation is to regulate emotions as a way to handle problems, while cognitive flexibility refers to the ability to handle more than one task at the same time. Cognitive emotion regulation can sometimes be maladaptive, and the effort of individuals might be unavailing. However, once cognitive flexibility is adopted and developed, it is expected to be adaptive in solving problems at the cognitive level. This chapter explores reciprocal associations among executive functions by mainly focusing on working memory, cognitive emotion regulation and cognitive flexibility. Further studies are advised to be conducted between cognitive emotion regulation strategies and working memory capacity, as these findings may have significant implications for understanding the correlation between memory and emotion. Cognitive flexibility is also advised to be researched more in order to understand its role in cognitive processes.

Keywords: working memory capacity, cognitive emotion regulation, cognitive flexibility, memory, acceptance

1. Introduction

The tendency to develop strategies for managing less familiar situations is a human-specific feature that can be capsuled within executive function set in the prefrontal cortex. The learning environment of individuals is one of those places in which unfamiliar or less familiar information is most likely to be encountered, and both executive functions and strategies are necessitated. Strategies for managing cognitive tasks can be needed in memory for recalling the necessary information and using it even when it is manipulated and also in cognitive emotion regulation for regulating negative emotions, as in reappraisal. The necessity for using strategies is also of great importance during the encoding of information, internalization and acquisition it. The common grounds of all these different components necessary for managing cognitive tasks are that they are related to executive functions such as attention, inhibition, working memory capacity, decision making, cognitive emotion regulation, cognitive flexibility, planning, language activities, reasoning, problem-solving and doing multiple cognitive functions simultaneously. These cognitive processes might allow one to address some claims that those higher-order executive functions might be possibly related to each other as well as being triggered reciprocally. Researchers focus on establishing an association among those cognitive processes with the aim of understanding how one handles these processes in mind naturally by using various techniques, research modes and approaches. As numerous studies have primarily focused on the possible associations of those cognitive processes, hereby this review aims to illustrate the association among some of them, which are working memory, cognitive emotion regulation and cognitive flexibility. In doing so, it was expected to discuss the importance of those executive functions' nature by designing such research.

Working Memory (WM) capacity is one of those higher-order executive function-related capacities that is defined as the capacity for handling the manipulation of information during cognitive activity, and it, as intellectual functioning, refers to a limited capacity in which necessary information is temporarily stored and manipulated simultaneously. A large body of research has focused on the WM capacity of individuals, and recent research in this vein suggests that individuals with high WM capacity are more accurate in cognitive tasks, and it is related to fluid intelligence, information compression, attention, cognitive flexibility all of which are executive functions of human cognition. Numerous studies have been conducted in the field of memory with the aim of understanding how inputs are processed in mind while encoding and recalling them and if their features are visuospatial, verbal or phonological matter. Some studies point to the fact that different components of working memory are most probably processed both in domain-general and domain-specific networks and put the domain-general network into a place that activates itself during working memory retention and visual and verbal working memory activation [1]. Neuropsychological evidences with regard to domain-related concerns supports that being specific to the domain might be related to the maintenance of the input rather than storing it. Working Memory (WM) is a nonunitary system and surely has subsystems that have its own storage capacity in which the input is retained before being recorded to the long-term memory (LTM) [2]. Recent neuropsychological evidences adopting the nonunitary system and the domain-general view supports the approach of Baddeley, the creator of Baddeley's Working Memory Model and contrasts with the domain-specific and unitary system approach by Cowan [3, 4].

In our recent research article, we brought forward the same question in a different design in which participants attended three complex span tasks measuring their working memory capacity from visual, verbal and spatial perspectives. The results of the present study indicated that scores obtained from the Symmetry Span Task, which is a spatial WM task, are correlated with the scores of the Rotation Span Task, which is a verbal and nonspatial task. The correlation between a spatial and verbal task indicates that individuals with a high WM score in Symmetry Span Task also score high in the Rotation Span Task. Studies conducted with individuals with psychological or neurological disorders mostly highlight the difference between spatial and verbal WM capacity. Taking an earlier study conducted with patients with Parkinson's disease having mild clinical symptoms as an example, it indicated that their spatial WM capacity was weaker when compared to verbal and visual WM capacity [5]. Also, a very recent study investigated the difference between verbal and visual WM capacity in depressive individuals and found that only the spatial WM capacity of individuals with unipolar depression and bipolar II depression was impaired [6]. There are some other researches indicating the difference between verbal and spatial WM capacity in healthy children. A study conducted with children indicated that verbal and spatial WM capacity have different neural bases in children [7]. Additionally, verbal WM capacity was found in a study as a good predictor for reading comprehension in children rather than spatial WM [8]. There are also other studies indicating the relationship between reading skills and verbal WM capacity [9, 10]. Another study questioning the role of age in WM capacity found that while verbal WM capacity is protected across years, the capacity for visual–spatial WM declines in time [11]. The results of the above-mentioned studies support the domain-specific view in the field of working memory. On top of all these studies pointing to the difference between verbal and spatial WM and supporting the domain-specific view, the present study conducted with healthy university students indicates a positive correlation between the visual and verbal WM capacity of individuals. Our study contrasts other research and supports the domain's general view. Similar research suggests a common storage for verbal, spatial or visuospatial WM as indicated in the present study [1, 12, 13]. When evaluated in terms of multitasking, the result of a very recent study researching the verbal and visuospatial WM capacity within a multitasking concept in which a Bayesian state-trace analysis was used indicated that while dealing with more than one task at the same time, a common pool for verbal and visuospatial data is used which supports the domain-general view [14]. Considering all of these studies, it can be speculated that domain-general and domain-specific tendencies may differ with regard to the scenario, cognitive load, age, psychological state, and other relevant indicators. The results that we endorse in this study support the domain's general view.

2. The working memory capacity and cognitive emotion regulation

Emerging studies reveal that any cognitive activity of high importance, such as decision-making, inhibition, language learning, etc., is connected with another cognitive behavior related to memory consolidation, evaluative thinking, and so on, addressing us to conceive that cognitive activities process within unitary systems. Cognitive emotion regulation is one of these important cognitive activities and has an untaught association with relevant cognitive processes. Additionally, the innate tendency of individuals to cope with emotions is instinctively shaped by their learned strategies, and these strategies are either adaptive or maladaptive strategies with which individuals tend to lower their stress levels arising from the acute problem. The role of memory herein in this process is open to question. Before discussing the role of memory, it might be of great importance to highlight that a growing body of literature with mounting evidence points to the memory systems, rather than a single memory pool. Recent approaches focus on the role of working memory in terms of the rationale that it is defined as having a connection with other subsystems and evaluated as a "capacity". Individuals with high working memory capacity, in its simplest definition, have the capacity to connect the information that is temporarily kept in the short-term memory to the relevant information in the long-term memory. By doing so, this capacity lessens the cognitive load. As the involvement of emotion during this cognitive process is inevitable, an adaptive regulation of it might also lessen the cognitive load, quicken the cognitive process, and contribute to the accurate processing of the cognitive regulation. Another rationale behind this prediction would be that if the cognitive process functions accurately in one system, then other systems can saliently function in an appropriate way, which is the exact regulation of activities and emotions. McRae & Gross [15] describe the role of emotion regulation within peoples' tendency to use strategies against discrepancies and contradictions in their emotional situation and the realities that they face. Being an unconscious process, cognitive emotion regulation can also be defined as the ability to function with the aim of balancing emotions. This process aims to lessen the stress level and get off the discrepancies. Accumulating studies conducted for researching the relation between emotion and memory revealed that people with high working memory capacity are prone to using more emotion regulation strategies [16], and working memory training contributes to the adaptive use of emotion regulation [17–19]. The results collected from these studies point to the fact that emotion and memory are associated, work reciprocally, and affect each other's functioning process as well as contributing to its accuracy if adopted adaptively. As cognitive emotion regulation and working memory capacity are predictors of some psychological disorders, such as depression, anxiety, and trauma-related disorders, this association must be evaluated within the perspectives of neuroplasticity, cognitive skills and associative memory, as some studies suggest [20].

3. The working memory capacity and cognitive flexibility

What working memory does is simply to switch between subsystems of memory while keeping the data to-be-encoded vivid. During this switching process, the working memory evidently works on the data and in the background, the attention is tried to be kept on it. This process is expressed as a capacity as it differs from person to person and is effected both by internal and external factors. Thus, the functions of working memory can be saliently framed in the fact that this process is a flexible and ongoing process as well as being open to the manipulation of these external and internal stimuli. In this regard, a possible association between working memory capacity and cognitive flexibility would reveal a scenario in which they act reciprocally. Cognitive flexibility can be defined as an ability with which more than one cognitive task can be handled, and it is one of the most prominent higher-order cognitive functions [21]. Explicitly linked to complex cognitive functions, it can be manipulated that this ability might decrease with aging as well as being affected by depression, anxiety and trauma like other cognitive functions [22]. Cognitive flexibility might be placed next to the definition of "task switching" in terms of pointing to its cognitive ground. Additionally, cognitive flexibility is a crucial facet in all branches of life as it necessitates being somehow skilled at finalizing a task, doing multiple tasks simultaneously without being distracted by the current task, and having psychological adjustment to adapt to the current situation. Its relation with working memory is still a question waiting to be answered. Some studies with healthy individuals have primarily focused on the relationship between Working Memory Capacity (WMC) and cognitive flexibility, and they indicated that those two cognitive processes predict creative thinking processes [23]. Other important issues might arise regarding how one controls these cognitive processes, which is highly advised to be studied. Working memory functions in association with multiple subsystems utilize various tasks in multiple concepts, and likewise, cognitive flexibility is related to the cognitive adaptation of an individual to various situations by doing multiple tasks. Thus, the important issue that arises here is that an advantageous working memory capacity might predict the cognitive flexibility of individuals. With the aim of understanding this possible reciprocal association, further studies must be carried out, and this logical rationale must be hypostasized. Further studies with larger samples might contribute to understanding how cognitive processes function and how much they are associative.

4. Discussing the reciprocal associations of cognitive functions

Our primary objective with this chapter is to delve into the logic behind the associations among working memory capacity, cognitive flexibility and the use of cognitive emotion regulation strategies, all of which have cognitive grounds. The results of our latest research have been elaborated in this review paper, which indicated that there is a significant difference and negative correlation between a working memory task, an Operation Span Task and an adaptive cognitive emotion regulation strategy that is Refocus on Planning. The findings of the research in question also indicated reciprocal correlations between Cognitive Emotion Regulation strategies. When findings collected from the Cognitive Flexibility scale were evaluated, they indicated a strong positive relationship between Refocus on Planning and Positive Reappraisal as cognitive emotion regulation strategies. Apart from this, Cognitive Flexibility was not found to be correlated with working memory capacity.

4.1 Understanding the relation between working memory capacity and cognitive emotion regulation strategies

Emotions are affective processes supporting or limiting cognitive abilities or vice versa supported or limited by cognitive processes, which can saliently highlight the fact that emotions find their expressions through cognitive processes as well as cognitive processes are not independent of emotions. It is next to impossible to evaluate emotions and cognitive processes separately, let alone cognitive emotion regulation strategies. Except that their regulation strategies and the coping mechanism that they adopt are adaptive or maladaptive, individuals have innate tendencies for coping with a problem with the aim of protecting the unity in their self, identity and memory. This tendency might be framed within an adaptive solution as in *"Refocus on Planning"*, or a maladaptive one as in *"Catastrophizing"*. It is definitely of a great certainty that regulating emotions is not limited to these adaptive or maladaptive cognitive emotion regulation strategies. Finding it unbearable or agonizing, some individuals might dissociate these feelings by suppressing them. On the other hand, individuals who are aware of the problem and conscious about what they are through but do not know how to cope with these problems and lose track of time and solutions by struggling desperately take a dip in maladaptive attitude.

With the aim of understanding the reciprocal relation between working memory capacity and the use of cognitive emotion regulation strategies, a comparison analysis was conducted in our latest research. The results collected from the Operation Span Task, a nonspatial and verbal task measuring the working memory capacity of an individual, illustrated that the WM capacity is significantly different from one of the adaptive cognitive emotion regulation strategies of individuals, that is the Refocus on Planning strategy and is strongly correlated with it. Refocus on Planning is an adaptive strategy with which individuals focus on possible solutions in order to manage their emotions [24]. On the other hand, WM capacity is robustly related to fluid intelligence [25–28], executive functions [29–31] and executive attention [32]. Focusing on the fact that Refocus on Planning is an adaptive strategy supporting the resilience in patients with depression and anxiety disorders [33], and this adaptive strategy is associated with more positive emotions because of the fact that individuals have lower anxiety symptoms when they design more plans for their problems [34], the low working memory capacity then necessitates more plans. In other words, the rationale behind this result would be that individuals with low working memory capacity might have the tendency of adopting more plans as well as refocusing on them as they cannot, on a balance of probabilities, rely on their working memory capacity which is a capacity of establishing an association between the subsystems of memory. Further studies must be conducted with regard to this association of working memory capacity and cognitive emotion regulation.

4.2 "Acceptance" as a cognitive emotion regulation strategy: is it adaptive or maladaptive?

Cognitive Emotion Regulation Strategies are adopted with the aim of managing a stressful or traumatic life event or a problem in order to protect the self. Adopting these strategies can end up either with self-injury, as in Self-blame or Catastrophizing, or by imposing its results on others, as in other-blame. Individuals might also feel lost in the loop of rumination, which is a destructive way of dealing with a problem and mostly ends with an increase in the level of stress. In other words, these are common maladaptive strategies for managing a problem that are not successful in relieving it, let alone finding a solution. Being antecedent-focused or response-focused, cognitive emotion regulation strategies can be adaptive, though. An adaptive way for individuals to deal with a problem might resemble like putting the Problem into Perspective, or Refocusing on Planning with the aim of handling a problem objectively, or employing Positive Reappraisal in order to see the contribution of the problem from a positive perspective. These strategies can be utilized together, and an adaptive strategy is expected to be associated with another adaptive strategy, rather than a maladaptive one. Our latest research questioned this associative relation, and the findings highlighted that cognitive emotion regulation strategies are correlated positively.

Taking Acceptance as an example, which is defined as accepting emotional experiences without manipulating them [35], it is contemplated as being correlated with positive matters [36]. Acceptance can be defined as an adaptive strategy as it can be saliently associated with consciousness while dealing with a stressful event, a contrary situation that might lead to psychopathology [37, 38] and anxiety disorders [38, 39]. Suppressed emotions or ideas can be dissociated when they are found unbearable. Contrarily, Acceptance might be more than dissociating by denying or suppressing it, but having a tendency for relieving stress. There are numerous studies illustrating that Acceptance contributes to reducing anxiety or depression [40], and healthy people prefer Acceptance as a strategy for regulating their emotions on a cognitive ground [41]. On the other hand, surprising results of some studies point to the association of Acceptance with negative emotions, depressive symptoms and self-injury [42, 43]. This contradiction raises the problem that the place of Acceptance is ambiguous. When the results of our latest research are formulated, they indicate that Acceptance,

Rumination, Self-blame and Catastrophizing function similarly. This result typically reveals that maladaptive cognitive emotion regulation strategies such as Rumination and Self-blame can be associated with an adaptive-looking maladaptive strategy, that is Acceptance. Rumination is the continuous contemplation of negative situations from negative perspectives. Rumination is employed by individuals diagnosed with Euthymic Bipolar Disorder and Major Depressive Disorder [44] and individuals with internalizing problems [45], and it has a role in mediating the relationship between anxiety problems and their consequent interference [46]. There are mounting studies pointing to the association of Rumination with Acceptance, other blame, and reduced Positive Reappraisal with maladaptive anger suppression [47]. Therefore, Acceptance is found to be associative either with adaptive or maladaptive coping strategies in various studies. This contradiction in results might allow us to address some claims that Acceptance is an adaptive-looking maladaptive strategy on the cognitive ground. Individuals having tendency to adopt Acceptance as a strategy for coping with a stressful event might increase their stress level as the nature of Acceptance might subconsciously evoke some feelings of being the victim of a sort of stressful situation as well as feeling less capable. These feelings are inevitably related to Self-blame. Another scenario would be that individuals accepting the situation as it is might find it difficult to resolve it, which is possibly why they directly accept it without any manipulation. Additionally, the result of our latest research revealed the relationship between Acceptance and Catastrophizing. Accumulating evidence from various research studies pointed out that Catastrophizing is associated with violent behaviors and somatic complaints [48, 49] and is mostly adopted by individuals diagnosed with euthymic bipolar disorder and Major Depressive Disorder [44]. This association also points to the same contradiction and allows us to address some claims that Acceptance can function as an adaptive strategy in some situations while some situations might turn it into a maladaptive one.

4.3 Reciprocal relationship among adaptive and maladaptive cognitive emotion regulation strategies

Individuals cope with problems in various ways by adopting different coping strategies. Strategies that contribute to resolving the problem as well as lowering the level of stress are defined as adaptive cognitive emotion regulation strategies. Our latest research focused on correlations among adaptive cognitive emotion regulation strategies and found them correlative. One adaptive coping strategy is Positive Refocusing, which finds its expression within the definition of focusing on the positive sides of the situation. There are various studies that have examined the contributions of Positive Refocusing, and they revealed that it is associated with externalizing problems and lowers the level of anxiety and depression [40, 45]. In our latest research, we found that Positive Refocusing is positively correlated with Refocus on Planning, which is saliently a cognitive strategy to handle stressful life events with resilience [50]. Numerous studies indicate the contribution of Positive Refocusing on the healthy and accurate mental process during regulating emotions [33, 43, 51]. What is more, Refocus on Planning was also found to be positively correlated with Positive Reappraisal, which is another adaptive strategy reducing the effect of negative events, such as bullying [52] and Putting into Perspective, an adaptive version of emotion regulation on the cognitive ground which is effective as an alternative to treatment [53]. Additionally, Positive Reappraisal and Putting into Perspective were also found correlative. Positive Reappraisal is the re-evaluation of the current stressful situation from a positive perspective, and a recent study revealed that high-trait-anxious women use the Positive Reappraisal strategy unsatisfactorily [54]. Our latest research found a negative correlation between Positive Reappraisal and Catastrophizing, a maladaptive strategy, which is mostly associated with negative feelings, violent behavior, somatic complaints, rumination, self-blame, other-blame, anxiety problems, and their consequent interference [44, 46, 48, 49]. The findings of our study contribute to the existing literature as it points to the nature of cognitive emotion regulation strategies, which are negatively correlated in terms of being adaptive and maladaptive.

When maladaptive cognitive emotion regulation strategies are evaluated, it can be explicitly acknowledged that they change reciprocally and predict each other. An individual tending to adopt a maladaptive strategy might be expected to utilize other maladaptive strategies with different scenarios. Thus, our latest research revealed positive correlations among maladaptive cognitive emotion regulation strategies. Self-blame and Rumination are maladaptive coping strategies associated with depressive modes, somatic complaints, anxiety problems and violent behavior [42, 46, 48, 49]. Numerous studies highlight the relationship between catastrophizing, rumination, self-blame and other-blame and anxiety problems, their consequent interference and problematic online gaming [46, 55]. When individuals are trained to gain perspectives in resolving a stressful problem, this association between maladaptive cognitive emotion regulation strategies might change them with adaptive ones. Therefore, in order to disrupt the domino effect of maladaptive strategies, cognitive emotion regulation training is highly recommended.

4.4 The nature of cognitive flexibility and its association with cognitive emotion regulation strategies

Characterized by being explicitly related to task switching, adapting to new changing situations, handling two or more tasks simultaneously, and focusing on the task while inhibiting irrelevant stimulus, cognitive flexibility is an ability of great importance for advancing cognitive processes and, apart from being aware of alternatives for a situation waiting to be resolved, it is also about being willing to adapt to this situation with a high self-efficacy [56]. In addition to these cognitive advantages, cognitive flexibility can also be an advantage in understanding emotions [57]. On the other hand, a possible relationship between cognitive flexibility and cognitive emotion regulation would be discussed in terms of adaptive strategies because of the nature of cognitive flexibility that is far from being maladaptive. In other words, cognitive flexibility is a quite adaptive cognitive process adopted with the aim of resolving a matter. As in cognitive flexibility, adaptive cognitive emotion regulation strategies help individuals manage their problems accurately. Our latest research reported that cognitive flexibility is associated and correlated with Refocus on Planning and Positive Reappraisal. Functioning as adaptive mechanisms, Refocus on Planning and Positive Reappraisal help individuals deal with a problem with lucidity by approaching the negative and stressful situation with a different perspective as well as evaluating possible positive aspects of it. Thus, the rationale behind the association of cognitive flexibility with two adaptive cognitive emotion regulation strategies would be that individuals with high levels of cognitive flexibility also innately adapt to multitasking in an adaptive way or vice versa; their cognitive flexibility skill might shape their way of handling a problem in an effective way which is possibly the reason why they use adaptive cognitive emotion regulation strategies.

DOI: http://dx.doi.org/10.5772/intechopen.1002893

5. Conclusion

Nature of working memory capacity can be capsulated within the definition of remembering and processing simultaneously. Its association with other cognitive processes finds its expression through the fact that it is a dynamic and active process reciprocally affecting other cognitive abilities, as in language learning [58, 59]. There is a growing body of literature questioning its association with other cognitive systems in various fields, such as cognitive psychology, neuropsychology as well as educational sciences. Studies conducted in this regard are performed by healthy individuals or those diagnosed with various psychological disorders. Therewithal, it should be acknowledged that an interdisciplinary perspective should be adopted for evaluating these explicit associations of each cognitive function with the other. For instance, a traumatic individual's approach to resolving a matter, remembering it, or deciding about it will definitely differ from the one who has not experienced a recent trauma. In a dissociated individual who becomes a subject with a psychiatric diagnosis due to a traumatic event that affects his/her consciousness, memory, identity and even autonomy because of individual and social traumas, when a recurrent flow of external traumatic stimuli evolves into a recurrent flow of internal traumatic stimuli, the emergence of the "traumatic self" is experienced [60]. This way, the traumatic self might be expected to be less skilled in resolving a stressful event, making a decision on it, thinking flexibly, regulating his/her emotions in an adaptive way as well as associating information in their short-term memory with their long-term memory, as working memory does. In other words, a trauma might have a domino effect on the cognitive processes. Trauma is only one small example explaining this effect. There are numerous studies pointing to the effect and association of psychological situations and cognitive abilities on each other. Individuals with low self-esteem employ avoidant and panic decision-making style, acute stress increases the possibility of decision-making failures and the functions of working memory and action-state orientation predicts self-control, all of which highlight this association [61–63]. Thus, an interdisciplinary perspective is advised to be adopted to discuss it.

What we questioned and researched in our latest study was the reciprocal association among some of these executive functions. The working memory capacity of individuals were determined with complex span tasks and the result was compared with their cognitive flexibility level, and tendencies for utilizing cognitive emotion regulation strategies. The overall result of the study indicated a relationship between working memory capacity and cognitive emotion regulation strategies. Another significant result was the association between cognitive flexibility and adaptive cognitive emotion regulation strategies. The findings of our research did not point to a relation between working memory capacity and cognitive flexibility. These results might provide a comprehensive understanding of the associations between cognitive processes. It should also be highlighted that individuals with a high level of cognitive flexibility can multitask, and this ability contributes to the regulation of their emotions, which is why cognitive flexibility training is another issue to be handled and developed for improving multitasking skills as well as regulating emotions.

Acknowledgements

We especially thank all our students in the Science Club of Istanbul Gelisim University and Istanbul Nisantasi University who volunteered their time to our study.

Conflict of interest

The authors declare no conflict of interest.

Author details

Kahraman Guler[1] and Aylin Aydin[2,3]*

1 Psychology Department, Dogus University, Istanbul, Turkey

2 Applied English and Translation, Istanbul Nisantasi University, Cerrahpasa, Institute of Forensic Sciences and Legal Medicine, Istanbul, Turkey

3 Institute of Forensic Sciences and Legal Medicine, Istanbul University - Cerrahpasa, Istanbul, Turkey

*Address all correspondence to: aylinaydinacademic@gmail.com; aylinaydin@ogr.iuc.edu.tr

References

[1] Li D, Christ SE, Cowan N. Domain-general and domain-specific functional networks in working memory. NeuroImage. 2014;**102**:646-656

[2] Martin RC, Rapp B, Purcell J. Domain-specific working memory: Perspectives from cognitive neuropsychology. In: Working Memory: State of the Science. Oxford, United Kingdom: Oxford University Press; 2020. pp. 235-287

[3] Baddeley A. Working Memory. New York: Oxford University Press; 1986

[4] Cowan N. Attention and Memory: An Integrated Framework. Oxford, United Kingdom: Oxford University Press; 1998

[5] Owen AM, Iddon JL, Hodges JR, Summers BA, Robbins TW. Spatial and non-spatial WM at different stages of Parkinson's disease. Neuropsychologia. 1997;**35**(4):519-532

[6] Li Z, Chen J, Feng Y, Zhong S, Tian S, Dai Z, et al. Differences in verbal and spatial WM in patients with bipolar II and unipolar depression: An MSI study. BMC Psychiatry. 2021;**21**(1):1-11

[7] Demir ÖE, Prado J, Booth JR. The differential role of verbal and spatial WM in the neural basis of arithmetic. Developmental Neuropsychology. 2014;**39**(6):440-458. DOI: 10.1080/87565641.2014.939182

[8] Oakhill J, Yuill N, Garnham A. The differential relations between verbal, numerical and spatial WM abilities and children's reading comprehension. International Electronic Journal of Elementary Education. 2011;**4**(1):83-106

[9] De Jong PF. Understanding normal and impaired reading development: A WM perspective. In: WM and Education. United States: Academic Press; 2006. pp. 33-60

[10] Carretti B, Borella E, Cornoldi C, De Beni R. Role of WM in explaining the performance of individuals with specific reading comprehension difficulties: A meta-analysis. Learning and Individual Differences. 2009;**19**(2):246-251

[11] Swanson HL. Verbal and visual-spatial WM: What develops over a life span? Developmental Psychology. 2017;**53**(5):971

[12] Jarrold C, Tam H, Baddeley AD, Harvey CE. How does processing affect storage in working memory tasks? Evidence for both domain-general and domain-specific effects. Journal of Experimental Psychology: Learning, Memory, and Cognition. 2011;**37**(3):688-705. DOI: 10.1037/a0022527

[13] Ashkenazi S, Rosenberg-Lee M, Metcalfe AW, Swigart AG, Menon V. Visuo–spatial working memory is an important source of domain-general vulnerability in the development of arithmetic cognition. Neuropsychologia. 2013;**51**(11):2305-2317

[14] Vergauwe E, von Bastian CC, Kostova R, Morey CC. Storage and processing in working memory: A single, domain-general resource explains multitasking. Journal of Experimental Psychology. General. 2022;**151**(2):285-301. DOI: 10.1037/xge0000895

[15] McRae K, Gross JJ. Emotion regulation. Emotion. 2020;**20**(1):1-9. DOI: 10.1037/emo0000703

[16] Jasielska A, Kaczmarek L, Brońska A, Dominiak M, Niemier K, Patalas D, et al. The relationship between WM and emotion regulation strategies. Roczniki Psychologiczne. 2015;**18**(4):567-578

[17] Xiu L, Zhou R, Jiang Y. WM training improves emotion regulation ability, evidence from HRV. Physiological Behavior. 2016;**155**:25-29. DOI: 10.1016/j.physbeh.2015.12.004

[18] Martin RC, Bridgett DJ, Mayes LC, Rutherford HJ. Maternal WM, emotion regulation, and responsivity to infant distress. Journal of Applied Developmental Psychology. 2020;**71**: 101202

[19] Deng Y, Hou L, Chen X, Zhou R. WM training improves emotion regulation in drug abstainers, evidence from frontal alpha asymmetry. Neuroscience Letters. 2021;**742**:135513

[20] Aydın S. Cross-validated adaboost classification of emotion regulation strategies identified by spectral coherence in resting-state. Neuroinformatics. 2022;**20**(3):627-639

[21] Magnusson KR, Brim BL. The aging brain. In: Brim BL, editor. Reference Module in Biomedical Sciences. Burlington, MA: Elsevier; 2014

[22] Miyake A, Friedman NP, Emerson MJ, Witzki AH, Howerter A, Wager TD. The unity and diversity of executive functions and their contributions to complex "frontal lobe" tasks, A latent variable analysis. Cognitive Psychology. 2000;**41**(1):49-100

[23] Figueroa I, Do WM. Capacity and Cognitive Flexibility Predict Creative Thinking and Production? United States: George Mason University; 2017

[24] Garnefski N, Van Den Kommer T, Kraaij V, Teerds J, Legerstee J, Onstein E. The relationship between cognitive emotion regulation strategies and emotional problems, comparison between a clinical and a non-clinical sample. European Journal of Personality. 2002;**16**(5):403-420. DOI: 10.1002/per.458

[25] Cochrane A, Green CS. Trajectories of performance change indicate multiple dissociable links between working memory and fluid intelligence. NPJ Science Learning. 2021;**6**:33. DOI: 10.1038/s41539-021-00111-w

[26] Kanerva K, Kalakoski V. The predictive utility of a WM span task depends on processing demand and the cognitive task. Applied Cognitive Psychology. 2016;**30**(5):681-690

[27] Lucidi A, Loaiza V, Camos V, Barrouillet P. Assessing WM capacity through time-constrained elementary activities. The Journal of General Psychology. 2014;**141**(2):98-112

[28] Unsworth N, Redick TS, Heitz RP, Broadway JM, Engle RW. Complex WM span tasks and higher-order cognition, a latent-variable analysis of the relationship between processing and storage. Memory. 2009;**17**(6):635-654

[29] Cartwright KB. Insights from cognitive neuroscience, The importance of executive function for early reading development and education. Early Education & Development. 2012;**23**(1):24-36

[30] Miyake A, Friedman NP. The nature and organization of individual differences in executive functions, four general conclusions. Current Directions in Psychological Science. 2012;**21**(1):8-14

[31] Willoughby MT, Kupersmidt JB, Voegler-Lee ME. Is preschool executive function causally related to academic achievement? Child Neuropsychology. 2012;**18**(1):79-91

[32] Unsworth N, Engle RW. The nature of individual differences in WM capacity, active maintenance in primary memory and controlled search from secondary memory. Psychological Review. 2007;**114**(1):104

[33] Min JA, Yu JJ, Lee CU, Chae JH. Cognitive emotion regulation strategies contributing to resilience in patients with depression and/or anxiety disorders. Comprehensive Psychiatry. 2013;**54**(8):1190-1197

[34] Sacchi L, Dan-Glauser E. Never too late to plan, "Refocus on planning" as an effective way to lower symptoms and difficulties in emotion regulation during the COVID-19 first lockdown. Emotion. 2021;**21**(7):1483

[35] Levitt JT, Brown TA, Orsillo SM, Barlow DH. The effects of acceptance versus suppression of emotion on subjective and psychophysiological response to carbon dioxide challenge in patients with panic disorder. Behavior Therapy. 2004;**35**(4):747-766

[36] Hayes SC, Strosahl KD, Wilson KG. Acceptance and Commitment Therapy: An Experiential Approach to Behavior Change. New York: Guilford Press; 1999

[37] Ehring T, Quack D. Emotion regulation difficulties in trauma survivors, The role of trauma type and PTSD symptom severity. Behavior Therapy. 2010;**41**(4):587-598

[38] Tull MT, Barrett HM, McMillan ES, Roemer L. A preliminary investigation of the relationship between emotion regulation difficulties and posttraumatic stress symptoms. Behavior Therapy. 2007;**38**(3):303-313

[39] McLaughlin KA, Mennin DS, Farach FJ. The contributory role of worry in emotion generation and dysregulation in generalized anxiety disorder. Behaviour Research and Therapy. 2007;**45**(8):1735-1752

[40] Wang QQ, Fang YY, Huang HL, Lv WJ, Wang XX, Yang TT, et al. Anxiety, depression, and cognitive emotion regulation strategies in Chinese nurses during the COVID-19 outbreak. Journal of Nursing Management. 2021;**29**(5):1263-1274

[41] Mazaheri M, Afshar H, Nikneshan S, Adibi P. Cognitive emotion regulation strategies in patients with functional dyspepsia and healthy controls: A comparative study. Advanced Biomedical Research. 2016;**2016**:5

[42] Lei H, Zhang X, Cai L, Wang Y, Bai M, Zhu X. Cognitive emotion regulation strategies in outpatients with major depressive disorder. Psychiatry Research. 2014;**218**(1-2):87-92

[43] Madjar N, Segal N, Eger G, Shoval G. Exploring particular facets of cognitive emotion regulation and their relationships with nonsuicidal self-injury among adolescents. Crisis. Jul 2019;**40**(4):280-286

[44] Wolkenstein L, Zwick JC, Hautzinger M, Joormann J. Cognitive emotion regulation in euthymic bipolar disorder. Journal of Affective Disorders. 2014;**160**:92-97

[45] Garnefski N, Kraaij V, van Etten M. Specificity of relations between adolescents' cognitive emotion regulation

strategies and internalizing and externalizing psychopathology. Journal of Adolescence. 2005;**28**(5):619-631

[46] Rodríguez-Menchón M, Orgilés M, Fernández-Martínez I, Espada JP, Morales A. Rumination, catastrophizing, and other-blame, the cognitive-emotional regulation strategies involved in anxiety-related life interference in anxious children. Child Psychiatry & Human Development. 2021;**52**(1):63-76

[47] Martin RC, Dahlen ER. Cognitive emotion regulation in the prediction of depression, anxiety, stress, and anger. Personality and Individual Differences. 2005;**39**(7):1249-1260. DOI: 10.1016/j.paid.2005.06.004

[48] Bao P, Jing J, Yang WH, Li XH, Cai YS. Violence-related behaviors among adolescents and its association with cognitive emotion regulation strategies. World Journal of Pediatrics. 2016;**12**(1):82-87

[49] Garnefski N, van Rood Y, De Roos C, Kraaij V. Relationships between traumatic life events, cognitive emotion regulation strategies, and somatic complaints. Journal of Clinical Psychology in Medical Settings. 2017;**24**(2):144-151

[50] Li L, Zhu X, Yang Y, He J, Yi J, Wang Y, et al. Cognitive emotion regulation, characteristics and effect on quality of life in women with breast cancer. Health and Quality of Life Outcomes. 2015;**13**(1):1-10

[51] Bernuzzi C, Setti I, Maffoni M, Sommovigo V. From moral distress to burnout through work-family conflict, the protective role of resilience and positive refocusing. Ethics & Behavior. 2021;**2021**:1-23

[52] Garnefski N, Kraaij V. Bully victimization and emotional problems in adolescents, moderation by specific cognitive coping strategies? Journal of Adolescence. 2014;**37**(7):1153-1160

[53] Wisco BE, Sloan DM, Marx BP. Cognitive emotion regulation and written exposure therapy for posttraumatic stress disorder. Clinical Psychological Science. 2013;**1**(4): 435-442

[54] Qi S, Basanovic J, Wang L, Xiang S, Hu W, Yi X. Regulation of negative emotions through positive reappraisal and distancing in high-trait-anxious women. Journal of Affective Disorders. 2020;**267**:191-202

[55] Kökönyei G, Kocsel N, Király O, Griffiths MD, Galambos A, Magi A, et al. The role of cognitive emotion regulation strategies in problem gaming among adolescents, a nationally representative survey study. Frontiers in Psychiatry. 2019;**10**:273

[56] Martin MM, Rubin RB. A new measure of cognitive flexibility. Psychological Reports. 1995;**76**:623-626

[57] Wang X, Liu X, Feng T. The continuous impact of cognitive flexibility on the development of emotion understanding in children aged 4 and 5 years: A longitudinal study. Journal of Experimental Child Psychology. 2021;**203**:105018

[58] Baddeley A. Working memory and language: An overview. Journal of Communication Disorders. 2003;**36**(3):189-208

[59] Ercetin G. Working memory and L2 reading: Theoretical and methodological issues. ELT Research Journal. 2015;**4**(2):101-110

[60] Ozturk E. Shared dissociative identity disorder and defector alter personality: Controlled human syndrome and the objectification trap phenomenon as a gaslighting form based on dissociative narcissism from the perspective of dissoanalysis theory and dissoanalytic psychohistory. Medicine. Science. 2023;**2023**:12

[61] Yesilbursa CC, Yalin FA. Sosyal Bilgiler Ogretmen Adaylarinin Karar Vermede Oz Saygi ve Karar Verme Stillerinin Incelenmesi. Turkish Studies-Educational Sciences. 2021;**16**(5):2512-2528

[62] Yilmaz S, Kafadar H. Decision-making under stress: Executive functions, analytical intelligence, somatic markers, and personality traits in young adults. Applied Neuropsychology: Adult. 2022;**2022**:1-15

[63] Gross D, Kohlmann CW. Predicting self-control capacity–taking into account working memory capacity, motivation, and heart rate variability. Acta Psychologica. 2020;**209**:103131

Section 2

Emotional Memory and Processes in Olfactory and Limbic Structures

Chapter 4

Cellular Processes and Synaptic Interactions in Nuclei of the Amygdala

Thomas Heinbockel

Abstract

The amygdala is a core structure of the limbic system in the brain. Anatomically, the amygdaloid complex comprises ~13 nuclei in the mid-temporal lobe. The amygdaloid complex is important for regulating emotional behavior, anxiety, fear, learning and memory. It is involved in several neurological disorders such as post-traumatic stress syndrome, depression, and temporal lobe epilepsy. The lateral nucleus of the amygdala is the main sensory input station of the amygdala and receives sensory information from cortical and subcortical (thalamic) fields. Thalamic afferents project to the lateral amygdala medially from the internal capsule, whereas cortical afferents arrive from the internal capsule. These two input pathways converge on populations of principal neurons and interneurons, both of which can be identified by characteristic electrophysiological, neurochemical, and morphological properties. Pressing issues for our understanding of the organization and operation of the amygdala are the functional significance of modulatory inputs from various signaling systems and the plasticity of its synaptic circuitry in relation to its pathway-specific inputs. This chapter reviews progress in this regard as far as cellular processes and synaptic interactions in nuclei of the amygdala are concerned which will help with our understanding of neural mechanisms underlying fear, anxiety, and related clinical disorders.

Keywords: amygdala, cortex, emotion, fear, intercalated cells, memory, synaptic plasticity, thalamus

1. Introduction

The amygdala, also known as amygdaloid complex or amygdaloid body, consists of several related nuclei. The amygdala is involved in different functional contexts such as memory, learning, emotion, fear, and motivation. Moreover, neurons in the amygdala contribute to symptoms of temporal lobe epilepsy and spread of seizure discharges in models of epilepsy. The amygdala is a core structure of the limbic system and a key site for emotion, anxiety, fear, learning and memory [1, 2]. Anatomically, the amygdaloid complex comprises ~13 nuclei in the midtemporal lobe. The amygdaloid complex is important for regulating emotional behavior and learning and is involved in several neurological disorders such as post-traumatic stress syndrome, depression, and temporal lobe epilepsy [3–5]. Pressing issues for our understanding of the organization and operation of the amygdala are the functional significance of modulatory inputs from various signaling systems and the plasticity of its synaptic circuitry in relation to its pathway-specific inputs. Recent studies set out to determine (a) the role of signaling molecules and their receptors for nerve cell signaling and information processing and (b) the influence of these signaling systems on learning and memory processes in the amygdala. The long-term objective of these studies is to understand the neural mechanisms underlying fear, anxiety, and related clinical disorders. The amygdala is critically involved in mediating normal and pathological conditions of fear and anxiety. Furthermore, the amygdala serves as a relay station for sensory input and output to higher order brain centers and centers of autonomic control. The lateral amygdaloid nucleus (lateral amygdala, LA) is the initial site of synaptic integration for cortical and thalamic afferent input to the amygdala. Signal flow of lateral amygdala excitatory principal neurons to downstream basolateral and central amygdaloid nuclei is influenced by two populations of GABAergic interneurons, classic local interneurons scattered throughout the amygdala and clusters of paracapsular intercalated (ITC) cells. These interneurons control impulse traffic through the amygdala, are sites of synaptic plasticity, and control fear memory. Using genetically engineered mice in which specific neuronal populations are labeled with fluorescence markers, it is possible to classify interneurons by describing their molecular, cellular and electrophysiological properties. Patch-clamp electrophysiology can be used to determine activity patterns and plastic behavior of principal neurons and interneurons in acute brain slices. Results obtained in these studies aid in better understanding the mechanisms of fear and anxiety, and consequently anxiety disorders, and might yield a new approach in fighting anxiety and stress disorders.

2. Fear, stress, and anxiety disorders

Fear and stress responses are facets of a behavioral repertoire that contributes to the survival of an individual. However, pathological deviations from normal fear and stress responses manifest themselves as clinical disorders including mood, fear, and anxiety disorders such as panic disorder and post-traumatic stress disorder (PTSD). PTSD has a lifetime prevalence of 6.8% of the adult population in the United States [6]. The amygdala is an anatomical and functional brain substrate critically involved in mediating normal and pathological conditions. Despite the clinical and functional importance of the amygdala, it is only recently that general principles of intra-amygdaloid mechanisms of signal processing have emerged from electrophysiological, anatomical, and pharmacological studies.

The amygdaloid complex is important for regulating emotional behavior and learning [7–13]. It is involved in several neurological disorders. Some of these disorders may also lie at the root of the many well-documented race-based physical health disparities that affect African Americans and other minority populations [14]. For example, after hurricane Katrina, emotional and stress-related health problems affected primarily members of minority populations, and stress through racism affects primarily black men.

3. Structural and functional organization of the lateral amygdala

The amygdala is a central relay station for sensory input and output to higher order brain centers and centers of autonomic control. The lateral nucleus of the amygdala is the initial site of synaptic integration for cortical and thalamic afferent input to the amygdala. Synaptic information is then relayed to the basolateral nucleus with the central nucleus serving as the amygdala output station. In rat brain slices, sharp-electrode recordings in an interface chamber can be used to determine physiological and synaptic parameters of lateral amygdala neurons [15, 16]. Neurons that are labeled intracellularly and digitally reconstructed allow a correlation between structure and function (**Figures 1** and 2). Principal neurons can be identified by

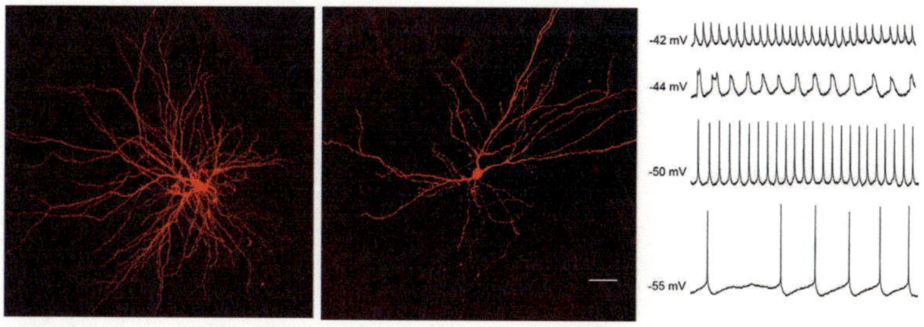

Figure 1.
Left and middle panels: confocal images of two principal neurons with many spine-rich dendrites in the lateral amygdala recorded intracellularly, stained with biocytin and incubated with Cy3-streptavidin. The neuron on the left has a pyramidal appearance, whereas the one in the center is more stellate in shape. Scale bar, for left image: 100 μm, center image: 150 μm. Right panel: intracellular recordings from a principal neuron in the lateral amygdala. Two types of oscillations of the membrane potential (low threshold [e.g., at −50 mV] and high threshold oscillations [at −42 mV and −44 mV]) are generated in principal neurons of the lateral amygdala.

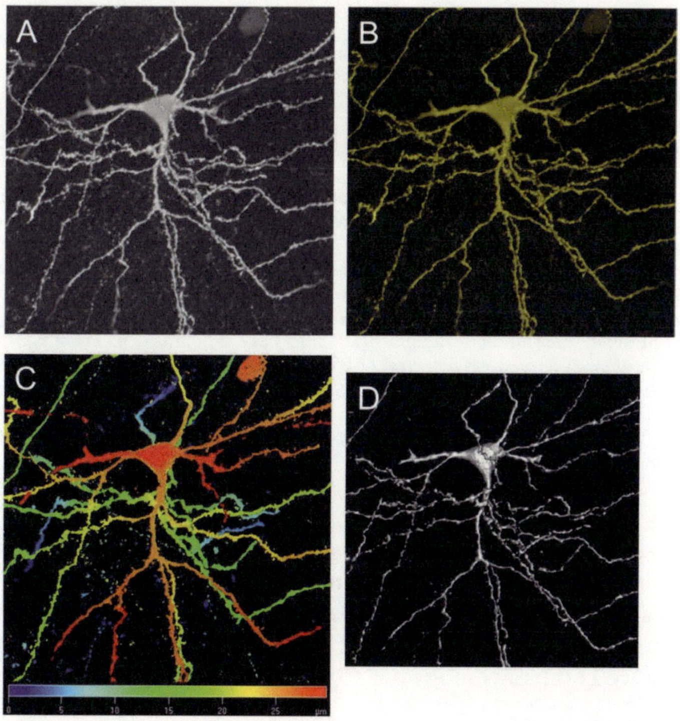

Figure 2.
Four images of the same principal neuron in the lateral amygdala, stained with biocytin and incubated with Cy3-streptavidin. (A) Laser scanning confocal image of the neuron. (B) False color image of the neuron. (C) Color-coded spatial image of the neuron. (D) Surface-rendered image of the neuron to generate 3-D effect.

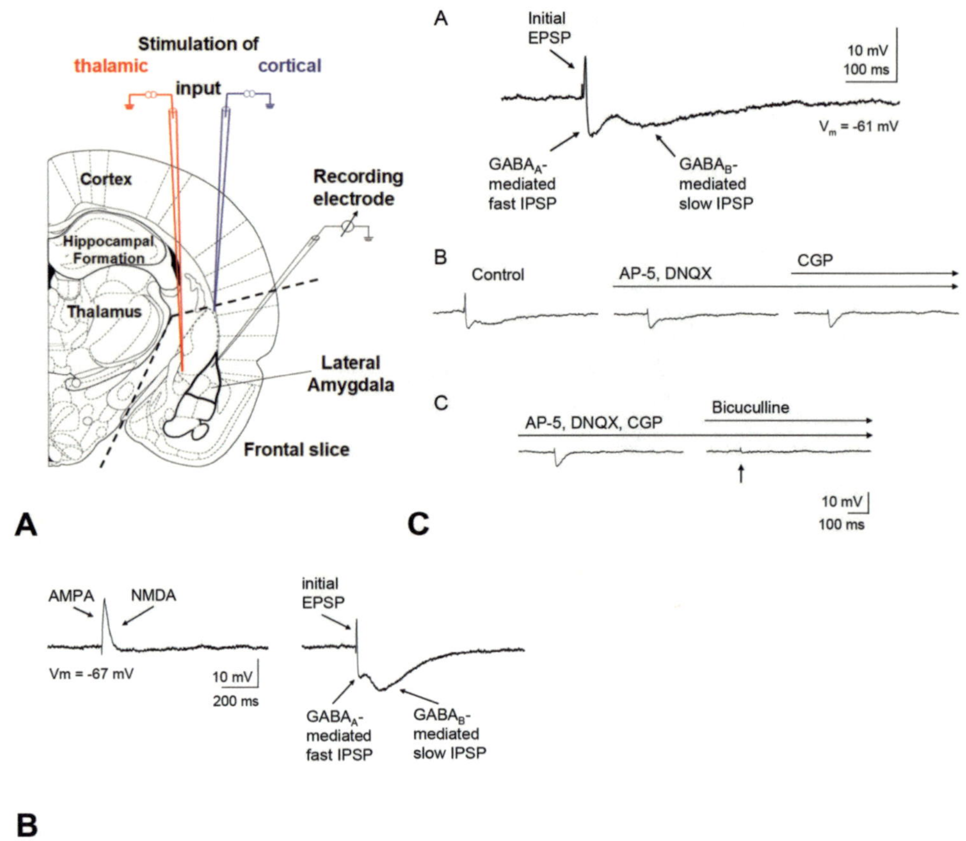

Figure 3.
(A) Amygdala slice preparation and activation of afferent input fibers to the lateral amygdala in vitro. Dashed lines indicate borders of the slice preparation. (B) Postsynaptic responses of principal neurons in the lateral amygdala evoked by electrical stimulation of thalamic or cortical inputs in fontal amygdaloid slices. Synaptic responses were evoked with two bipolar tungsten electrodes placed in the external capsule and close to the internal capsule for stimulation of putative cortical and thalamic afferents. Stimulus intensity was adjusted to produce a synaptic response 30–50% of maximum amplitude without triggering action potentials. Neurons were held at resting membrane potential. A single principal neuron in the lateral amygdala receives input from both pathways, cortical and thalamic. Single electrical stimulation of inputs evokes either a primarily excitatory response mediated by ionotropic glutamate receptors (left trace) or a triphasic sequence of a fast glutamate receptor mediated excitatory postsynaptic potential (EPSP) followed by a fast $GABA_A$ and a slow $GABA_B$ receptor mediated inhibitory postsynaptic potential (IPSP) (right trace). (C) The fast IPSP in principal neurons is mediated by GABA release from interneurons and can be pharmacologically isolated with NMDA (AP5) and non-NMDA (DNQX) antagonists and a $GABA_B$ antagonist (CGP35348). Bicuculline blocks the fast $GABA_A$-mediated IPSP. Modified from [15].

morphological (after Biocytin injection and histological processing) and electrophysiological criteria. Projection neurons reveal complex morphologies with numerous spiny dendrites (**Figures 1** and **2**). Principal neurons in the amygdala generate slow membrane potential oscillations [1, 17]. Specifically, neurons produce theta activity (6–10 Hz) (**Figure 1**) during emotional arousal and various types of rhythmic activity during sleep.

An acute brain slice preparation that contains the amygdala and adjacent regions of the brain, e.g., cortical and thalamic areas (**Figure 3**) is an advantageous experimental system for studies of amygdala signaling [15, 16]. This slice preparation preserves the synaptic circuitry in the amygdala to the extent that one can study synaptic processing by electrically stimulating afferent input (stimulation electrodes) while recording from neurons (recording electrode) in the lateral amygdala or other amygdaloid nuclei. It is possible to record and intracellularly stain principal neurons and interneurons in the amygdala while selectively activating thalamic or cortical input fibers (**Figure 3**). Thereby, one can synaptically drive postsynaptic neurons and induce neuroplasticity in an input-specific manner. Such frontal amygdaloid slices, prepared from deeply anesthetized rats of either sex (postnatal days 25–30) [16], allow to evoke synaptic responses from principal neurons in the lateral amygdala with two bipolar tungsten electrodes placed in the external capsule and close to the internal capsule for stimulation of putative cortical and thalamic afferents, respectively [18, 19]. Stimulus intensity can be adjusted to produce a synaptic response 30–50% of maximum amplitude without triggering action potentials. In such experiments, neurons are held at resting membrane potential.

Recordings from principal neurons in the lateral amygdala have confirmed that electrical stimulation of afferents that project to the lateral amygdala evokes excitatory and inhibitory postsynaptic potentials (PSPs) and thus results in a dual effect with direct excitation of principal neurons coupled with concurrent feedforward inhibition via interneurons [17, 20–22]. The excitatory PSPs (EPSPs) are mediated by feedforward glutamate receptor-mediated excitation. The inhibitory PSPs (IPSPs) are either polysynaptic in origin or they are monosynaptically mediated by feedforward inhibition via local GABAergic interneurons since (a) the IPSPs can be evoked after blocking excitatory transmission (**Figure 3**) [15], (b) connections between different basolateral amygdala nuclei primarily consist of excitatory connections [23], and (c) lesions deafferenting the basolateral complex lead to minor decreases in glutamic acid decarboxylase levels [24]. Thus, the excitability of principal neurons in the lateral amygdala depends upon the relative strength of the inputs to projection neurons and interneurons.

4. Synaptic plasticity and fear learning in the lateral amygdala

The lateral amygdala is the main sensory input station of the amygdala and receives sensory information from cortical and subcortical (thalamic) fields [25]. Thalamic afferents project to the lateral amygdala medially from the internal capsule, whereas cortical afferents arrive from the external capsule [26]. These two input pathways converge on principal neurons and interneurons, both of which have characteristic electrophysiological, neurochemical, and morphological properties [20, 21, 27–30]. Theta rhythm activity (**Figure 1**) has been implicated in lateral amygdala synaptic plasticity, retrieval of fear memory, and transmission in the amygdalo-hippocampal network [31]. Therefore, theta frequency stimulation can be used as a paradigm to test for intercellular signaling and synaptic plasticity in the lateral amygdala. We have reported that homosynaptic long-term depression (LTD) of excitatory responses can be induced in the lateral amygdala after stimulation of putative thalamic inputs but not of cortical inputs (**Figure 4**) [16]. This form of LTD is induced by theta frequency stimulation and involves postsynaptic Ca^{2+}-dependent mechanisms and metabotropic glutamate receptors. The reader is referred to [2] for an excellent review of plastic synaptic networks of the amygdala for the acquisition, expression, and extinction of conditioned fear.

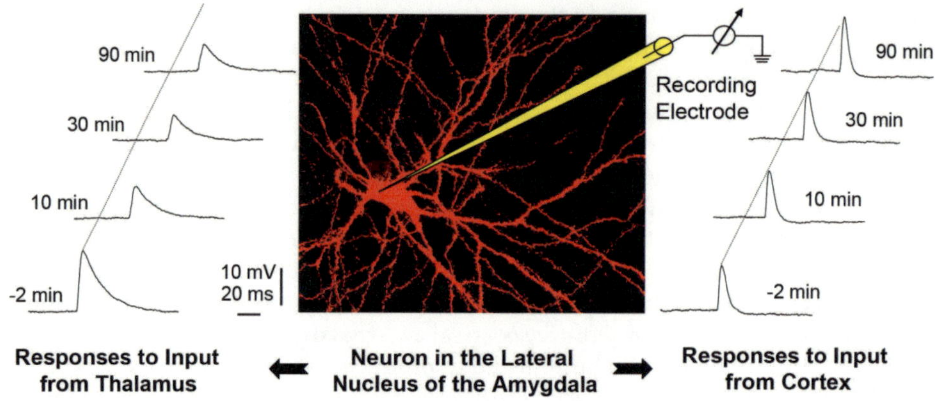

Figure 4.

Theta frequency stimulation of thalamic afferents evokes an LTD of EPSP amplitude evoked with single thalamic control stimuli while it has no effect on cortically evoked EPSPs. Theta stimulation of cortical input fibers does not result in a change of EPSP amplitude of either afferent input pathway demonstrating input specific plasticity, namely, depression of EPSPs in response to stimulation of thalamic afferents. Modified from [16].

5. Principles of signal processing in the amygdala

Three principles of signal processing have emerged from studies of amygdala circuitry [1, 2]: (1) oscillatory activity in the theta frequency range, (2) dominance of inhibition and (3) sustained synaptic plasticity. One principle of information processing in the amygdala is the stereotypic responses in the form of slow oscillating activity in the theta frequency range [1, 32]. Studies of the membrane properties of projection neurons in the lateral amygdala revealed that these neurons are capable of generating intrinsic stable oscillations in the frequency range of 6–10 Hz (**Figure 1**) which is reminiscent of theta waves in the limbic system. The frequency of the oscillations is relatively constant even with excitatory synaptic input of different strength. The functional consequence is a synchronization of intrinsic and synaptic input in this defined frequency and a temporal structuring of synaptic input activity toward slow rhythmic output [33, 34]. These oscillations form the basis of phase-coupled oscillatory response patterns that are evoked in neuronal networks of the amygdala. In amygdalohippocampal circuits, rhythmic synchronization at theta frequencies increases between the hippocampal CA1 region and the lateral amygdala after fear conditioning and presentation of fear stimuli [31]. Therefore, synchronized theta activities in the amygdalo-hippocampal network serve as a neuronal correlate of conditioned fear.

A second principle of information processing is the dominance of synaptic inhibition in the amygdala (**Figure 3**). Mainly feedforward inhibitory responses are observed to stimulation of major afferent systems through activation of GABAergic or glycinergic receptors [21, 22, 35–38]. Powerful IPSPs regulate the responses of projection neurons in the lateral amygdala relatively irrespective of the stimulation site, i.e., perirhinal, entorhinal, basomedial or lateral amygdala stimulation [35]. Thus, the lateral amygdala seems to be equipped with an inhibitory gating mechanism regulating information flow through the amygdala [27, 39]. We found that adenosine may participate in these neuronal processes related to fear conditioning, learning and memory in the amygdala [15], e.g., by directly interacting with the inhibitory mechanisms and/or by modulating the potentiation of $GABA_A$-mediated synaptic currents

DOI: http://dx.doi.org/10.5772/intechopen.1005123

in pyramidal neurons after tetanic stimulation of inputs to interneurons [28]. A decrease of GABAergic interneuron activity in the lateral amygdala is correlated with the development of epileptiform discharges. In the kindling model of epilepsy, GABA receptor-mediated inhibitory transmission is reduced [40–42] and glutamatergic transmission is enhanced in the amygdala [41, 42].

Another principle of processing in the lateral amygdala is the prevalence of synaptic plasticity, including input-specific neural plasticity of its afferent pathways (**Figure 4**). Different forms of activity-dependent synaptic plasticity have been reported to occur in the amygdala, one of which, long-term potentiation, has been suggested to be an important cellular mechanism for conditioned fear [9]. During fear conditioning the synapses of thalamic afferent pathways in the lateral amygdala undergo synaptic plasticity [43, 44] illustrating the relevance of associative LTP for learning processes in a behavioral context. The molecular mechanisms of LTP have been studied as well as other forms of plasticity such as LTD and bidirectional synaptic plasticity (depression and potentiation) that is sensitive to the type and sequence of the stimulation paradigm used [16, 45–47]. Modulation of synaptic transmission in the amygdala by activity-dependent processes is controversially discussed in the literature, and it is not yet clear how the different input systems relate to synaptic plasticity in the amygdala and to conditioned fear. Short-term and long-term synaptic plasticity in the amygdala appear to play a critical role in conditioned fear and anxiety.

6. Synaptic signaling mechanisms and plasticity in amygdala interneurons

Fear learning involves the lateral and basolateral nuclei (BLA) of the amygdala. Here the association between incoming fearful and neutral stimuli leads to potentiation of synaptic transmission. Projections from the basolateral amygdala nucleus synapse on neurons in the central nucleus which in turn projects its axons to the hypothalamus and brainstem to trigger the autonomic expression of fear [9]. However, this classic view was recently challenged [48] in favor of a model in which afferent input reaches both the lateral and central amygdala. Lateral amygdala neurons do not directly synapse on central amygdala output neurons, but rather lateral amygdala neurons are thought to synapse on interneurons which in turn disinhibit brainstem projecting central amygdala neurons. The central amygdala itself can be a site of synaptic plasticity. The GABAergic interneurons of the paracapsular intercalated (ITC) cell masses that surround the lateral amygdala and basolateral amygdala have a pivotal role in mediating signal flow from lateral to central amygdala [39, 49–53]. Therefore, pressing issues for understanding the organization and operation of the amygdala are the functional significance of modulatory input (e.g., [54, 55]), the functional role of its distinct neuronal populations, and the plasticity of its synaptic circuitry in relation to its pathway-specific inputs.

One of the emerging principles of synaptic processing in the amygdala is the dominance of inhibition. In addition to excitatory principal neurons, the amygdala houses two populations of GABAergic interneurons: (1) classic local interneurons scattered throughout the amygdala and (2) clusters of paracapsular intercalated (ITC) cells (**Figure 5**). Afferent synaptic input converges onto single principal neurons as well as interneurons in the lateral amygdala. The relay from lateral amygdala principal neurons to downstream amygdala nuclei is regulated by GABAergic interneurons through feedforward and feedback inhibition. Classic interneurons in the lateral amygdala respond with EPSCs of constant latencies to minimal stimulation of both cortical and

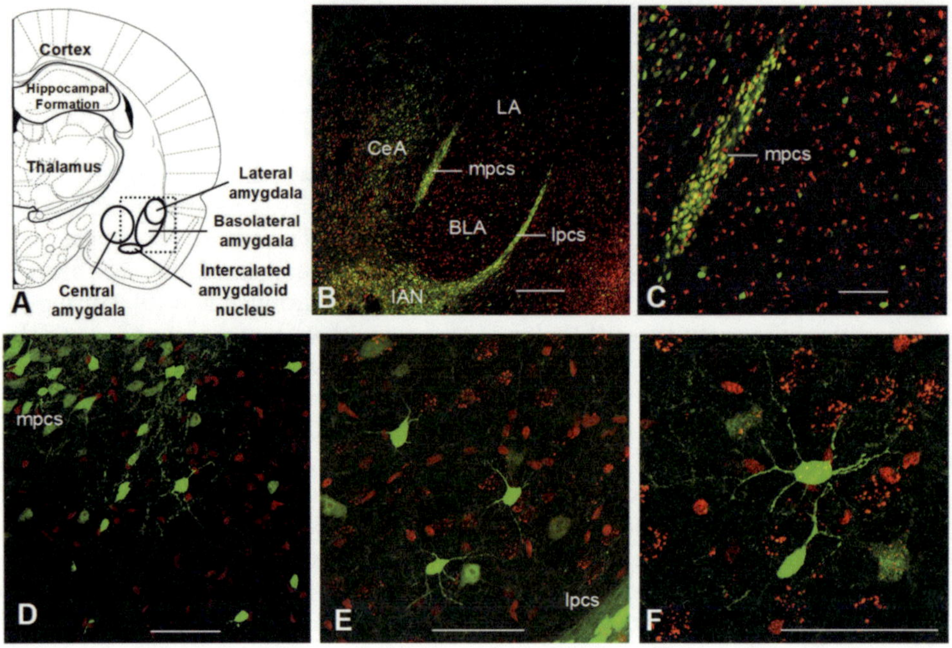

Figure 5.
Confocal images of amygdala slices from GAD-67-GFP mice counterstained with a nuclear stain (Sytox orange). (A) Schematic diagram of parts of the amygdala. Dotted rectangle indicates outlines of panel B. (B) Paracapsular intercalated cells form an inhibitory sheath surrounding the basolateral amygdala (BLA). Clusters of intercalated cells are located primarily along the external capsule (lateral paracapsular cells, lpcs) and along the border of BLA and central amygdala (CeA) (medial pcs, mpcs). Scattered GABAergic interneurons are found throughout the BLA and adjacent areas. The intercalated nucleus is located immediately ventro-medial to the BLA. (C) The medial cluster of paracapsular cells contains many GABAergic INs; same slice as in B. (D) Higher magnification of parts of the medial paracapsular cell cluster and scattered INs in the BLA; same slice as in B. (E) Scatterd INs and parts of the lateral paracapsular cell cluster, same slice as in B. (F) Dendritic branching pattern of an IN in BLA; same slice as in B. Scale bars: B—200 μm, C—100 μm, D—50 μm, E—50 μm, F—50 μm. BLA—basolateral amygdala; CeA—central amygdala; IAN—intercalated amygdaloid nucleus; LA—lateral amygdala; lpcs—lateral paracapsular cells; mpcs—medial paracapsular cells.

thalamic fibers, indicating direct convergent monosynaptic input [29]. Based on the data, a population of interneurons in the lateral amygdala is likely to receive convergent input from thalamic and cortical fibers through principal neurons both directly and indirectly (in a feedforward and feedback manner, respectively). Irrespective of the route of excitation, interneurons in the lateral amygdala can be regarded as homogeneous with respect to their afferent connections considering thalamic and cortical input pathways and mediate inhibitory control of postsynaptic principal neurons. This symmetrically built GABAergic circuitry can be of functional significance, given the distinctive role of the two afferent input systems for the mediation of different components of fear responses and the importance of GABAergic mechanisms for limitation of excessive neuronal activity [1, 29].

7. Intercalated cells of the amygdala

Another principle of processing in the lateral amygdala is the prevalence of synaptic plasticity, including input-specific neural plasticity of its afferent pathways. Based on the findings discussed above, one hypothesis is that Pavlovian fear conditioning and fear memory involve synaptic plasticity also of the inhibitory GABAergic interneurons. Specifically, (a) activation of interneurons in the lateral amygdala might regulate impulse traffic of its principal neurons, and (b) interneurons in the amygdala are sites of input-specific and location-specific neural plasticity themselves. These interneurons include scattered interneurons in the amygdala and the paracapsular ITC neurons [56]. Genetically engineered mice, GAD-67-GFP mice [57, 58], house specific neuronal populations, namely GABAergic interneurons, that are labeled with fluorescence markers to identify and classify GABAergic neurons in slice preparations *in vitro*, based upon patch-clamp recordings, molecular and cellular markers (**Figure 5**). The nomenclature for ITC cell masses is confusing. While some authors refer to all intercalated cell masses as intercalated nuclei, others restrict the term nucleus to a prominent one that is filled with inhibitory neurons and is located immediately ventral to the basolateral amygdala. Here, this nucleus is referred to as the intercalated amygdaloid nucleus whereas the clusters of GABAergic interneurons that form a sheath around the basolateral amygdala complex are referred to as paracapsular ITC cell masses (**Figure 5**). We distinguish two clusters, a lateral one along the external capsule (lateral paracapsular cells, lpcs) and a medial one along the border of the BLA and CeA (medial pcs, mpcs) in accordance with a recent study of ITC cells in the amygdala [59].

In the GAD-67-GFP mouse line, GABAergic interneurons are fluorescently labeled with GFP such that GABAergic cells in slices from these mice can be directly visualized with fluorescence microscopy in the patch-clamp slice microscope. In this transgenic mouse line, the promoter for GAD-67 drives the expression of GFP. In addition to their fluorescent property, amygdala interneurons can be recognized by electrophysiological criteria. Classic interneurons respond with regular sustained firing to depolarizing current pulses, they possess high resting input resistance and generate fast spikes compared with those of principal cells [17, 27–29, 60, 61]. Interneurons typically have non-spiny or sparsely spiny dendrites (**Figure 5**).

Few studies have addressed synaptic signaling mechanisms and plasticity in amygdala interneurons. After tetanic electrical stimulation of the external capsule, The authors in [28] observed long-term potentiation (LTP) mediated by calcium-permeable AMPA receptors in putative GABAergic interneurons in the basolateral amygdala. This LTP is reflected in an increased $GABA_A$-mediated inhibitory synaptic current in principal neurons. In another study, tetanization of thalamic afferents resulted in NMDA-dependent heterosynaptic LTP of excitatory postsynaptic currents (EPSCs) in interneurons which in turn was reflected as larger IPSCs in postsynaptic principal neurons [62]. However, the potentiation was found in both input pathways, thalamic and cortical. Interneurons of ITC cell masses located at the border between central amygdala and basolateral amygdala (putative medial paracapsular cells, mpcs) displayed bidirectional synaptic plasticity (LTD, LTP) depending on the stimulus paradigm used [63, 64]. More recent evidence points to a role of ITC cell masses in fear extinction [65, 66] and a specific role of neuropeptide S in interneuron-related fear processes [65, 67]. ITC cells form a sheath of GABAergic interneurons surrounding the basolateral amygdala [2, 68]. Additionally, a prominent intercalated nucleus of GABAergic cells is located ventro-medially of the basolateral (**Figure 5**). Intercalated cells of the medial group, which is located at the border of basolateral amygdala and central amygdala, display distinct short-term plastic synaptic responses that vary dramatically between different connected cell pairs [69]. Intercalated cells exhibit an unusual pattern of receptor expression and do not stain for typical interneuron markers such as calbindin, calretinin, parvalbumin, NPY, CCK, somatostatin [59].

Furthermore, only study has attempted to systematically analyze the large number of anatomically distinct ITC interneurons in terms of electrophysiological parameters [59]. Overall, these studies reveal an exciting, yet inconclusive picture of interneuron function in the amygdala and do not provide a clear functional assignment and/or classification of GABAergic cells. The data available suggest that, as in other neural systems [70], interneurons in the amygdala present themselves as diverse types of neurons [30, 71] and possibly participate in spatiotemporal division of labor. This idea is supported by two studies of amygdala neural circuits [54, 72]. Interneurons in the amygdala are found in specific locations and patterns of aggregation, either in ITC clusters or distributed as individual interneurons throughout amygdaloid nuclei. These interneurons can also be distinguished based on morphological features. In confocal images, interneurons of ITC cell masses have smaller soma size, and their dendritic branching is less readily visible than the more extensive branching of scattered interneurons. This suggests that differences in dendritic receptive fields exist that in turn could result in fewer synaptic connections of ITC cells and a possible dominance of short-range synaptic interactions with neighboring ITC cells. Thus, one possible role of ITC cells could be to mediate lateral inhibition in ITC cell clusters whereas classic interneurons with longer dendrites could function in longer range feedforward or feedback inhibition of principal neurons. Furthermore, afferent input from thalamic and cortical regions can either converge on individual interneurons and principal neurons or reach amygdaloid neurons in a more distributed manner [29]. All of this suggests that subpopulations of interneurons exhibit distinct functional roles in amygdaloid networks. The basis of these functional roles rests with distinct intrinsic and synaptic properties of interneurons which are supported by two studies that showed how distinct neuronal circuits can differentially affect amygdala function [54, 72].

8. Fear conditioning, functional imaging, and theta rhythms

Functional imaging studies, such as fMRI (functional magnetic resonance imaging) studies, have confirmed the central role of the amygdala in fear conditioning paradigms. Fear conditioning is a classical experimental paradigm used to investigate associative learning, particularly the formation and expression of fear-related memories. In this paradigm, participants are exposed to neutral stimuli (such as tones or images) paired with aversive stimuli (such as electric shocks or unpleasant images). Over time, the neutral stimuli become associated with fear responses [73–75]. The amygdala is critically involved in the acquisition and consolidation of fear-related memories during fear conditioning. Studies have consistently shown that activity in the amygdala increases when animal models (rodents) or humans are exposed to conditioned stimuli associated with fear-inducing events. This increased activity reflects the encoding and storage of fear memories [4, 5, 9, 76, 77]. In fear conditioning paradigms, the amygdala also plays a role in the expression of fear responses. Once fear memories are formed, the amygdala becomes activated when individuals encounter the conditioned stimuli, even in the absence of the aversive stimuli. This activation leads to the expression of fear responses, such as increased arousal, physiological changes, and behavioral reactions. Functional neuroimaging techniques such as fMRI examine the neural correlates of fear conditioning in humans. Studies using fMRI have consistently implicated the amygdala in the processing of fear-related stimuli during fear conditioning tasks. Increased amygdala activity is observed when participants are exposed to conditioned stimuli associated with fear, providing evidence for its involvement in fear learning and expression [77–80]. Beyond its role as a primary hub for fear processing, the amygdala interacts with other brain regions to modulate fear responses. fMRI studies have revealed functional connectivity between the amygdala and various cortical and subcortical regions involved in emotional regulation, attention, and memory. These network interactions contribute to the complex processing of fear-related information and the regulation of emotional responses [77–80].

Fear conditioning represents a most reproducible paradigm to study the neurobiological mechanisms underlying emotions [81]. Studies of the synaptic plasticity underlying fear conditioning in both humans and rodent models focus on neural circuits in the amygdala and prefrontal cortex, cingulate gyrus, and hippocampus [82, 83]. Plastic changes involve cortical representations and the refinement of contextual input as part of circuit modulation. Theta oscillatory activity across several brain regions has been found to be critical in this regard [31, 84–87], e.g., in relation to long-term fear memory, modulation of synchronized oscillations in the hippocampal-prefrontal-amygdala circuit in contextual fear, and changes in brain rhythms and connectivity tracking fear acquisition and reversal. Projection neurons of the lateral amygdala exhibit resonant/oscillatory behavior of their membrane potential and provide a cellular correlate of coherent theta activity in amygdalo-hippocampal pathways (see **Figure 1**). This can be a neural correlate of conditioned fear [84, 85]. Synaptic plasticity observed in these networks has been proposed to rely on correlated activity, e.g., input-specific long-term depression of thalamo-amygdaloid signals and consolidation of long-term potentiation in the dentate gyrus. Moreover, increased theta rhythm in the cingulate cortex was found to subserve fear acquisition and is transmitted to other cortical regions via increased functional connectivity, which allows for fast theta rhythm synchronization [87].

9. Conclusions

A long-term objective of research in the amygdala is to understand the neural mechanisms underlying fear, anxiety, and related clinical disorders. The amygdala is critically involved in mediating normal and pathological conditions of fear and anxiety. The amygdala serves as a relay station for sensory input and output to higher order brain centers and centers of autonomic control. The lateral amygdaloid nucleus is the initial site of synaptic integration for cortical and thalamic afferent input to the amygdala. Signal flow of lateral amygdala excitatory principal neurons to downstream basolateral and central amygdaloid nuclei is influenced by two populations of GABAergic interneurons, classic local interneurons scattered throughout the amygdala and clusters of paracapsular intercalated cells. Interneurons control impulse traffic through the amygdala, are sites of synaptic plasticity, and control fear memory.

A basic principle of information processing in the amygdala is the stereotypic responses in the form of oscillating activity in the theta frequency range. Input specific plastic changes seen with the theta frequency stimulation paradigm make them extremely relevant for synaptic plasticity and help to integrate the intrinsic properties of the projection neurons into synaptic and network properties in the amygdala. Presumably, synaptic depression in the lateral amygdala, like other forms of enduring synaptic weakening, provides a means of reducing the relative contribution of a high synaptic input. Changes in synaptic strength are input specific and could mediate balance shifts of thalamic vs. cortical influences, possibly with effects on fear conditioning neuronal circuits. The functional consequence could be a reduction of reflex-like components of conditioned fear responses, the direct thalamic influences, with a shift toward a stronger cortical component which is linked to analysis of contextual stimuli. These studies determine the functional significance of neurotransmitter systems and provide insights into amygdala network properties. Furthermore, the results could help to explain and suggest treatments for pain, anxiety, phobias, stress, and neurological conditions associated with amygdala networks.

Acknowledgements

This work was supported in part by the National Science Foundation [NSF IOS-1355034], Howard University College of Medicine, and the District of Columbia Center for AIDS Research, an NIH funded program [P30AI117970], which is supported by the following NIH Co-Funding and Participating Institutes and Centers: NIAID, NCI, NICHD, NHLBI, NIDA, NIMH, NIA, NIDDK, NIMHD, NIDCR, NINR, FIC and OAR. The content is solely the responsibility of the author and does not necessarily represent the official views of the NIH.

I wish to express my sincere gratitude to Prof. Dr. Hans-Christian Pape for his guidance and support during my studies of the amygdala in his institute at the Otto-von-Guericke University, Magdeburg, Germany. I am grateful to Dr. Michael T. Shipley and Dr. Adam Puche (both University of Maryland School of Medicine, Baltimore, Maryland) for sharing GAD67-GFP labeled brain material. I would like to thank Prof. Dr. Yuchio Yanagawa (Gunma University, Gunma, Japan) for allowing me to use the GAD67-GFP knock-in mice that he generated. I wish to acknowledge the Integrated Imaging Center, Johns Hopkins University (Baltimore, Maryland) for using the facility, including the Zeiss LSM 510 VIS/Confocor 3 FCS.

Conflict of interest

The author declares no conflict of interest.

Author details

Thomas Heinbockel
Department of Anatomy, Howard University College of Medicine, Washington, DC, USA

*Address all correspondence to: theinbockel@howard.edu

References

[1] Pape H-C, Driesang RB, Heinbockel T, Laxmi TR, Meis S, Seidenbecher T, et al. Cellular processes in the amygdala: Gates to emotional memory? Zoology. 2001;**104**:232-240

[2] Pape HC, Pare D. Plastic synaptic networks of the amygdala for the acquisition, expression, and extinction of conditioned fear. Physiological Reviews. 2010;**90**(2):419-463. DOI: 10.1152/physrev.00037.2009. Erratum in: Physiological Reviews 2010;**90**(3):1269

[3] Johansen JP, Cain CK, Ostroff LE, LeDoux JE. Molecular mechanisms of fear learning and memory. Cell. 2011;**147**:509-524

[4] LeDoux JE. Rethinking the emotional brain. Neuron. 2012;**73**:653-676

[5] LeDoux JE. Coming to terms with fear. PNAS USA. 2014;**111**:2871-2878

[6] NIH website: Post-Traumatic Stress Disorder (PTSD). Available from: https://www.nimh.nih.gov/health/statistics/post-traumatic-stress-disorder-ptsd [Accessed: January 02, 2024]

[7] Davis M. Role of NMDA receptors and MAP kinase in the amygdala in extinction of fear: Clinical implications for exposure therapy. The European Journal of Neuroscience. 2002;**16**:395-398

[8] Herry C, Ciocchi S, Senn V, Demmou L, Muller C, Luthi A. Switching on and off fear by distinct neuronal circuits. Nature. 2008;**454**:600-606

[9] LeDoux JE. Emotion circuits in the brain. Annual Review of Neuroscience. 2000;**23**:155-184

[10] Levine J, Chengappa KN, Gershon S, Drevets W. Differentiating primary pathophysiologic from secondary adaptational processes. Depression and Anxiety. 2001;**14**:105-101

[11] Phelps EA, LeDoux JE. Contributions of the amygdala to emotion processing: From animal models to human behavior. Neuron. 2005;**48**(2):175-187. DOI: 10.1016/j.neuron.2005.09.025

[12] Ressler KJ, Mayberg HS. Targeting abnormal neural circuits in mood and anxiety disorders: From the laboratory to the clinic. Nature Neuroscience. 2007;**10**(9):1116-1124. DOI: 10.1038/nn1944

[13] Shin LM, Rauch SL, Pitman RK. Amygdala, medial prefrontal cortex, and hippocampal function in PTSD. Annals of the New York Academy of Sciences. 2006;**1071**:67-79

[14] Mays VM, Cochran SD, Barnes NW. Race, race-based discrimination, and health outcomes among African Americans. Annual Review of Psychology. 2007;**58**:201-225

[15] Heinbockel T, Pape H-C. Modulatory effects of adenosine on postsynaptic potentials of projection neurons in the lateral amygdala of the rat. British Journal of Pharmacology. 1999;**128**:190-196

[16] Heinbockel T, Pape H-C. Input-specific long-term depression in the lateral amygdala evoked by theta frequency stimulation. The Journal of Neuroscience. 2000;**20**(RC68):1-5

[17] Washburn MS, Moises HC. Electrophysiological and morphological properties of rat basolateral amygdaloid neurons in vitro. The Journal of Neuroscience. 1992;**12**:4066-4079

[18] Mahanty NK, Sah P. Excitatory synaptic inputs to pyramidal neurons of the lateral amygdala. The European Journal of Neuroscience. 1999;**11**:1217-1222

[19] Weisskopf MG, LeDoux JE. Distinct populations of NMDA receptors at subcortical and cortical inputs to principal cells of the lateral amygdala. Journal of Neurophysiology. 1999;**81**:930-934

[20] Rainnie DG, Asprodini EK, Shinnick-Gallagher P. Excitatory transmission in the basolateral amygdala. Journal of Neurophysiology. 1991;**66**:986-998

[21] Rainnie DG, Asprodini EK, Shinnick-Gallagher P. Inhibitory transmission in the basolateral amygdala. Journal of Neurophysiology. 1991;**66**:999-1009

[22] Washburn MS, Moises HC. Inhibitory responses of rat basolateral amygdaloid neurons recorded in vitro. Neuroscience. 1992;**50**:811-830

[23] Smith Y, Paré D. Intra-amygdaloid projections of the lateral nucleus in the cat: PHA-L anterograde labeling combined with postembedding GABA and glutamate immunocytochemistry. The Journal of Comparative Neurology. 1994;**342**(2):232-248. DOI: 10.1002/cne.903420207

[24] Le Gal LSG, Paxinos G, Ben-Ari Y. Neurochemical mapping of GABAergic systems in the amygdaloid complex and bed nucleus of the stria terminalis. Brain Research. 1978;**155**(2):397-403. DOI: 10.1016/0006-8993(78)91037-5

[25] Pitkanen A, Savander V, LeDoux JE. Organization of intra-amygdaloid circuitries in the rat: An emerging framework for understanding functions of the amygdala. Trends in Neurosciences. 1998;**20**:517-523

[26] LeDoux JE, Farb CR, Milner TA. Ultrastructure and synaptic associations of auditory thalamo-amygdala projections in the rat. Experimental Brain Research. 1991;**85**:577-586

[27] Lang EJ, Paré D. Synaptic responsiveness of interneurons of the cat lateral amygdaloid nucleus. Neuroscience. 1998;**83**(3):877-889. DOI: 10.1016/s0306-4522(97)00420-x

[28] Mahanty NK, Sah P. Calcium-permeable AMPA receptors mediate long-term potentiation in interneurons in the amygdala. Nature. 1998;**394**:683-687

[29] Szinyei C, Heinbockel T, Montagne J, Pape HC. Putative cortical and thalamic inputs elicit convergent excitation in a population of GABAergic interneurons of the lateral amygdala. The Journal of Neuroscience. 2000;**20**(23):8909-8915. DOI: 10.1523/JNEUROSCI.20-23-08909.2000

[30] Sah P, Faber ESL, Lopez de Armentia M, Power J. The amygdaloid complex: Anatomy and physiology. Physiological Reviews. 2003;**83**:803-834

[31] Seidenbecher T, Laxmi TR, Stork O, Pape H-C. Amygdalar and hippocampal theta rhythm synchronization during fear memory retrieval. Science. 2003;**301**:846-850

[32] Paré D, Pape H-C, Dong J. Bursting and oscillating neurons of the cat basolateral amygdaloid complex in vivo: Electrophysiological properties and morphological features. Journal of Neurophysiology. 1995;**74**:1179-1191

[33] Collins DR, Paré D. Reciprocal changes in the firing probability of lateral and central medial amygdala

neurons. The Journal of Neuroscience. 1999;**19**:836-844

[34] Paré D, Collins DR, Pelletier JG. Amygdala oscillations and the consolidation of emotional memories. Trends in Cognitive Sciences. 2002;**6**:306-314

[35] Lang EJ, Paré D. Similar inhibitory processes dominate the responses of cat lateral amygdaloid projection neurons to their various afferents. Journal of Neurophysiology. 1997;77(1):341-352. DOI: 10.1152/jn.1997.77.1.341

[36] Nose I, Higashi H, Inokuchi H, Nishi S. Synaptic responses of Guinea pig and rat central amygdala neurons in vitro. Journal of Neurophysiology. 1991;**65**:1227-1241

[37] Danober L, Pape H-C. Mechanisms and functional significance of a slow inhibitory potential in neurons of the lateral amygdala. The European Journal of Neuroscience. 1998;**10**:853-867

[38] Danober L, Pape H-C. Stychnine-sensitive glycine responses in neurons of the lateral amygdala: An electrophysiological and immunocytochemical characterization. Neuroscience. 1998;**85**:427-441

[39] Royer S, Martina M, Paré D. An inhibitory interface gates impulse traffic between the input and output stations of the amygdala. The Journal of Neuroscience. 1999;**19**(23):10575-10583. DOI: 10.1523/JNEUROSCI.19-23-10575.1999

[40] Gean PW, Shinnick-Gallagher P, Anderson AC. Spontaneous epileptiform activity and alteration of GABA- and of NMDA-mediated neurotransmission in amygdala neurons kindled in vivo. Brain Research. 1989;**494**:177-181

[41] Asprodini EK, Rainnie DG, Shinnick-Gallagher P. Epileptogenesis reduces the sensitivity of presynaptic gamma-aminobutyric acidB receptors on glutamatergic afferents in the amygdala. The Journal of Pharmacology and Experimental Therapeutics. 1992;**262**:1011-1021

[42] Rainnie DG, Asprodini EK, Shinnick-Gallagher P. Kindling-induced long-lasting changes in synaptic transmission in the basolateral amygdala. Journal of Neurophysiology. 1992;**67**:443-454

[43] McKernan MG, Shinnick-Gallagher P. Fear conditioning induces a lasting potentiation of synaptic currents in vitro. Nature. 1997;**390**(6660):607-611. DOI: 10.1038/37605

[44] Rogan MT, Stäubli UV, LeDoux JE. Fear conditioning induces associative long-term potentiation in the amygdala. Nature. 1997;**390**(6660):604-607. DOI: 10.1038/37601. Erratum in: Nature 1998;**391**(6669):818

[45] Huang YY, Kandel ER. Postsynaptic induction and PKA-dependent expression of LTP in the lateral amygdala. Neuron. 1998;**21**:169-178

[46] Huang YY, Kandel ER. Low-frequency stimulation induces a pathway-specific late phase of LTP in the amygdala that is mediated by PKA and dependent on protein synthesis. Learning & Memory. 2007;**14**:497-503

[47] Li H, Weiss SR, Chuang DM, Post RM, Rogawski MA. Bidirectional synaptic plasticity in the rat basolateral amygdala: Characterization of an activity-dependent switch sensitive to the presynaptic metabotropic glutamate receptor antagonist 2S-alpha-ethylglutamic acid. The Journal of Neuroscience. 1998;**18**:1662-1670

[48] Paré D, Quirk GJ, Ledoux JE. New vistas on amygdala networks in conditioned fear. Journal of Neurophysiology. 2004;**92**(1):1-9. DOI: 10.1152/jn.00153.2004

[49] Royer S, Martina M, Paré D. Polarized synaptic interactions between intercalated neurons of the amygdala. Journal of Neurophysiology. 2000;**83**:3509-3518

[50] Asede D, Bosch D, Lüthi A, Ferraguti F, Ehrlich I. Sensory inputs to intercalated cells provide fear-learning modulated inhibition to the basolateral amygdala. Neuron. 2015;**86**(2):541-554. DOI: 10.1016/j.neuron.2015.03.008. Epub 2015 Apr 2

[51] Asede D, Doddapaneni D, Bolton MM. Amygdala intercalated cells: Gate keepers and conveyors of internal state to the circuits of emotion. The Journal of Neuroscience. 2022;**42**(49):9098-9109. DOI: 10.1523/JNEUROSCI.1176-22.2022

[52] Hagihara KM, Bukalo O, Zeller M, Aksoy-Aksel A, Karalis N, Limoges A, et al. Intercalated amygdala clusters orchestrate a switch in fear state. Nature. 2021;**594**(7863):403-407. DOI: 10.1038/s41586-021-03593-1. Epub 2021 May 26

[53] Stern DB, Wilke A, Root CM. Anatomical connectivity of the intercalated cells of the amygdala. eNeuro. 2023;**10**(10):ENEURO.0238-23.2023. DOI: 10.1523/ENEURO.0238-23.2023

[54] Huber D, Veinante P, Stoop R. Vasopressin and oxytocin excite distinct neuronal populations in the central amygdala. Science. 2005;**308**(5719):245-248. DOI: 10.1126/science.1105636

[55] Neugebauer V, Mazzitelli M, Cragg B, Ji G, Navratilova E, Porreca F. Amygdala, neuropeptides, and chronic pain-related affective behaviors. Neuropharmacology. 2020;**170**:108052. DOI: 10.1016/j.neuropharm.2020.108052. Epub 2020 Mar 15

[56] Paré D, Smith Y. GABAergic projection from the intercalated cell masses of the amygdala to the basal forebrain in cats. The Journal of Comparative Neurology. 1994;**344**(1):33-49. DOI: 10.1002/cne.903440104

[57] Tamamaki N, Yanagawa Y, Tomioka R, Miyazaki J, Obata K, Kaneko T. Green fluorescent protein expression and colocalization with calretinin, parvalbumin, and somatostatin in the GAD67-GFP knock-in mouse. The Journal of Comparative Neurology. 2003;**467**(1):60-79. DOI: 10.1002/cne.10905

[58] Ono M, Yanagawa Y, Koyano K. GABAergic neurons in inferior colliculus of the GAD67-GFP knock-in mouse: Electrophysiological and morphological properties. Neuroscience Research. 2005;**51**(4):475-492. DOI: 10.1016/j.neures.2004.12.019

[59] Marowsky A, Yanagawa Y, Obata K, Vogt KE. A specialized subclass of interneurons mediates dopaminergic facilitation of amygdala function. Neuron. 2005;**48**(6):1025-1037. DOI: 10.1016/j.neuron.2005.10.029

[60] Sugita S, Tanaka E, North RA. Membrane properties and synaptic potentials of three types of neurone in rat lateral amygdala. Journal of Physiology (London). 1993;**460**:705-718

[61] Danober L, Heinbockel T, Driesang RB, Pape H-C. Synaptic mechanisms of NMDA-mediated hyperpolarization in lateral amygdaloid neurons. Neuroreport. 2000;**11**:2501-2506

[62] Bauer EP, LeDoux JE. Heterosynaptic long-term potentiation of inhibitory interneurons in the lateral amygdala. The Journal of Neuroscience. 2004;**24**:9507-9512

[63] Royer S, Paré D. Bidirectional synaptic plasticity in intercalated amygdala neurons and the extinction of conditioned fear responses. Neuroscience. 2002;**115**:455-462

[64] Royer S, Paré D. Conservation of total synaptic weight through balanced synaptic depression and potentiation. Nature. 2003;**422**:518-522

[65] Jüngling K, Seidenbecher T, Sosulina L, Lesting J, Sangha S, Clark SD, et al. Neuropeptide S-mediated control of fear expression and extinction: Role of intercalated GABAergic neurons in the amygdala. Neuron. 2008;**59**:298-310

[66] Likhtik E, Popa D, Apergis-Schoute J, Fidacaro GA Jr, Pare D. Amygdala intercalated neurons are required for expression of fear extinction. Nature. 2008;**454**:642-645

[67] Meis S, Bergado-Acosta JR, Yanagawa Y, Obata K, Stork O, Munsch T. Identification of a neuropeptide S responsive circuitry shaping amygdala activity via the endopiriform nucleus. PLoS One. 2008;**3**:e2695

[68] Millhouse OE. The intercalated cells of the amygdala. The Journal of Comparative Neurology. 1986;**247**(2):246-271. DOI: 10.1002/cne.902470209

[69] Geracitano R, Kaufmann WA, Szabo G, Ferraguti F, Capogna M. Synaptic heterogeneity between mouse paracapsular intercalated neurons of the amygdala. The Journal of Physiology. 2007;**585**:117-134

[70] Klausberger T, Somogyi P. Neuronal diversity and temporal dynamics: The unity of hippocampal circuit operations. Science. 2008;**321**:53-57

[71] Sosulina L, Meis S, Seifert G, Steinhauser C, Pape H-C. Classification of projection neurons and interneurons in the rat lateral amygdala based upon cluster analysis. Molecular and Cellular Neurosciences. 2006;**33**:57-67

[72] Herry C, Ciocchi S, Senn V, Demmou L, Müller C, Lüthi A. Switching on and off fear by distinct neuronal circuits. Nature. 2008;**454**(7204):600-606. DOI: 10.1038/nature07166. Epub 2008 Jul 9

[73] Luchkina NV, Bolshakov VY. Mechanisms of fear learning and extinction: Synaptic plasticity-fear memory connection. Psychopharmacology. 2019;**236**(1):163-182. DOI: 10.1007/s00213-018-5104-4. Epub 2018 Nov 10

[74] Meulders A. Fear in the context of pain: Lessons learned from 100 years of fear conditioning research. Behaviour Research and Therapy. 2020;**131**:103635. DOI: 10.1016/j.brat.2020.103635. Epub 2020 Apr 30

[75] Labrenz F, Woud ML, Elsenbruch S, Icenhour A. The good, the bad, and the ugly-chances, challenges, and clinical implications of avoidance research in psychosomatic medicine. Frontiers in Psychiatry. 2022;**13**:841734. DOI: 10.3389/fpsyt.2022.841734

[76] Meisner OC, Nair A, Chang SWC. Amygdala connectivity and implications for social cognition and disorders. Handbook of Clinical Neurology. 2022;**187**:381-403. DOI: 10.1016/B978-0-12-823493-8.00017-1

[77] Alvarez RP, Biggs A, Chen G, Pine DS, Grillon C. Contextual fear

conditioning in humans: Cortical-hippocampal and amygdala contributions. The Journal of Neuroscience. 2008;**28**(24):6211-6219. DOI: 10.1523/JNEUROSCI.1246-08.2008

[78] Geng H, Xu P, Aleman A, Qin S, Luo YJ. Dynamic organization of large-scale functional brain networks supports interactions between emotion and executive control. Neuroscience Bulletin. 2024. DOI: 10.1007/s12264-023-01168-w

[79] Tassone VK, Gholamali Nezhad F, Demchenko I, Rueda A, Bhat V. Amygdala biomarkers of treatment response in major depressive disorder: An fMRI systematic review of SSRI antidepressants. Psychiatry Research: Neuroimaging. 2024;**338**:111777. DOI: 10.1016/j.pscychresns.2023.111777. Epub 2024 Jan 3

[80] Battaglia S, Di Fazio C, Mazzà M, Tamietto M, Avenanti A. Targeting human glucocorticoid receptors in fear learning: A multiscale integrated approach to study functional connectivity. International Journal of Molecular Sciences. 2024;**25**(2):864. DOI: 10.3390/ijms25020864

[81] Traina G, Tuszynski JA. The neurotransmission basis of Post-traumatic stress disorders by the fear conditioning paradigm. International Journal of Molecular Sciences. 2023;**24**(22):16327. DOI: 10.3390/ijms242216327

[82] Ritger AC, Parker RK, Trask S, Ferrara NC. Elevated fear states facilitate ventral hippocampal engagement of basolateral amygdala neuronal activity. Frontiers in Behavioral Neuroscience. 2024;**18**:1347525. DOI: 10.3389/fnbeh.2024.1347525

[83] Plas SL, Tuna T, Bayer H, Juliano VAL, Sweck SO, Arellano Perez AD, et al. Neural circuits for the adaptive regulation of fear and extinction memory. Frontiers in Behavioral Neuroscience. 2024;**18**:1352797. DOI: 10.3389/fnbeh.2024.1352797

[84] Pape HC, Stork O. Genes and mechanisms in the amygdala involved in the formation of fear memory. Annals of the New York Academy of Sciences. 2003;**985**:92-105. DOI: 10.1111/j.1749-6632.2003.tb07074.x

[85] Pape HC, Narayanan RT, Smid J, Stork O, Seidenbecher T. Theta activity in neurons and networks of the amygdala related to long-term fear memory. Hippocampus. 2005;**15**(7):874-880. DOI: 10.1002/hipo.20120

[86] Stubbendorff C, Hale E, Bast T, Cassaday HJ, Martin SJ, Suwansawang S, et al. Dopamine D1-like receptors modulate synchronized oscillations in the hippocampal-prefrontal-amygdala circuit in contextual fear. Scientific Reports. 2023;**13**(1):17631. DOI: 10.1038/s41598-023-44772-6

[87] Pirazzini G, Starita F, Ricci G, Garofalo S, di Pellegrino G, Magosso E, et al. Changes in brain rhythms and connectivity tracking fear acquisition and reversal. Brain Structure & Function. 2023;**228**(5):1259-1281. DOI: 10.1007/s00429-023-02646-7. Epub 2023 May 2

Chapter 5

The Amygdaloid Body as the Anatomical Substrate of Emotional Memory: Implications in Health and Disease

Alessandro Weiss and Francesco Weiss

Abstract

The Amygdaloid Body is a heterogeneous nuclear complex that establishes extensive connections with numerous structures of the limbic system, the thalamus, the brainstem, and the neocortex, and constitutes the focal center of its widespread three-dimensional white matter chassis. Since the 50s, the neurophysiological observations of Wilder Penfield et al. began to clarify the role of the AB in human memory. More recently, the introductions of a more advanced neuroimaging technology (PET, fMRI, DTI) led to a growing awareness of its crucial implications in the etiology of a variety of neuropsychiatric disorders, such as trauma spectrum and mood spectrum disorders. Additionally, the AB and its connections have been successfully used as a target for Deep Brain Stimulation (DBS) in the treatment of refractory forms of psychiatric disorders, especially trauma spectrum disorders. Therefore, gaining a deeper understanding of the morphophysiology of the AB has increasingly become utmost relevance for neuroscientists and clinicians alike. With the present chapter, we attempt to provide an exhaustive description of the functional anatomy of the AB, hopefully providing a useful tool for the approach to the anatomical substrates of the emotional components of memory and learning and to their role in the phenomenology and treatment of neuropsychiatric disorders.

Keywords: emotional memory, amygdaloid body, post-traumatic stress disorder, bipolar disorder, implicit memory, explicit memory, memory, disconnection syndrome

1. Introduction

"Learning is defined as the change in behaviour that results from acquiring knowledge about the world, and memory as the process by which knowledge is encoded, stored, and later retrieved" [1].

Debate on the mind's mechanism for learning and memory dates back to the beginnings of western science when the ancient Greek philosophers speculated about the causes of behavior and the relation between mind and brain. In the seventeenth century, René Descartes believed in a dualistic view between body and mind. According to his view, the brain (part of the body) is responsible for perception, motor acts, memory, appetites, and passions, while mind for the conscious experiences characteristic of the human behavior. The connection between body and mind was supposed to be in the pineal gland, located in the center of the brain. In the eighteenth century, the philosophical debate about learning and memory split along different lines: Empiricists believed that the brain was initially a blank slate (tabula-rasa) that is later filled by sensory experiences, whereas idealists, notably Immanuel Kant, believed that the perception of the world was determined by inherent features (a priori) of human mind [1].

During the twentieth century, Freud and colleagues introduced the psychoanalytical thinking, which consisted in a renewed dualistic model of memory and learning based on conscious and unconscious processes [2]. Despite the importance of this revolutionary intuition, its biological confirmation had to wait for almost one century later when the development of modern neuroscience, open to the possibility of reconsidering previous philosophical thoughts, embraced a more pragmatic model based on functional and molecular neuroanatomy [1, 3–5]. The first observation to have been confirmed was properly the intuition of Descartes that there would be a higher-order matter (soul) controlling a lower one (body): since the second half of the twentieth century, it has been assumed that phylogenetic recent areas of the brain, disposed cranially in the neuraxis, control older areas disposed more caudally. On this respect, Paul MacLean in 1960 presented the model of the "triune brain," comprising a reptilian central primitive brain, devoted to homeostasis and survival; a paleomammalian brain, surrounding the central reptilian part, which is related to emotional and motivational aspects of behavior; and lastly, a neomammalian brain, corresponding to the neocortex, which is responsible for cognitive and executive integrations of behavior [6]. More recently, Heimer and Wilson [7] demonstrated that basal ganglia extend all the way to the ventral surface of the mammalian brain to include the olfactory tubercle and other structures of the basal proencephalon. They grouped together these portions of the basal ganglia into what they termed ventral striato-pallidal system. Such model replaced the old concept of the subcommissural substantia innominata, providing a more defined morphophysiologic establishment to these structures. The importance of this acquisition is not only due to more refined topographic depiction of the basal ganglia, but also and foremost to the understanding of the biunivocal relationship between the basal ganglia and the whole cortical mantle.

Contemporarily, the use of neural electrical stimulation in awake-neurosurgery, employed in the treatment of refractory temporal lobe epilepsy, allowed to clarify the anatomical substrate of memory and learning [1, 8–10]: it was demonstrated that mnesic functions were not grouped in a single region but widely distributed among many brain regions, whose access was independent of visual, verbal, or other sensory clues. The current classification of memory (shown in **Figure 1**) was elaborated by Peter Graf and Daniel Schacter, who based their model on the observation of post-operative mnesic impairments [1, 10, 11]. The authors classified mnesic functions in a short-term type and in a long-term type. Long-term memory, in turn, has been subdivided into implicit and explicit forms, differing in that the former does not require conscious awareness for recalling. Four different processes compose the physiologic build-up process of long-term memory: encoding, storage, consolidation, and retrieval [12].

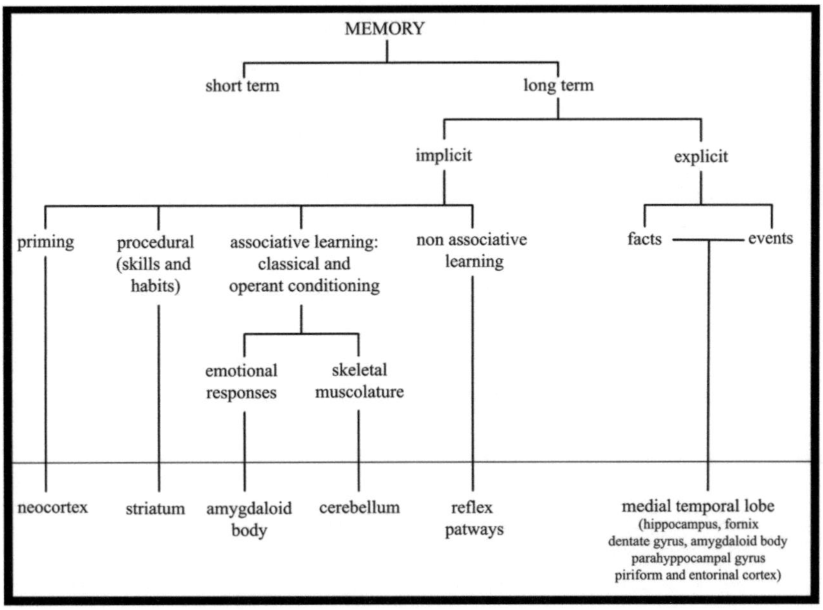

Figure 1.
Classification of memory by Graf and Schacter.

Implicit memory (also referred to as non-declarative or procedural memory) is unconscious, mainly automatic and tightly connected to the original conditions under which the learning occurred. It gives rise to priming, skill learning, habit memory, and conditioning and comprises the emotional and viscerosomatic memories.

Explicit memory (or declarative memory) is defined as the deliberate or conscious retrieval of previous experiences, as well as conscious recall of factual knowledge about people, places, and things and it can be further subdivided into an episodic or autobiographical memory (memory of personal experiences) and a semantic memory (memory of meanings, words, and concepts).

But why do humans store some memories and discard others? This question has persisted since the beginning of the philosophical debate on memory and learning. Nowadays, the intuitive assumption that memory consolidation and storage depend, to a substantial extent, on the affective charge of experiences is well established. Such function is largely under the control of the Amygdaloid Body (AB) [13]. Indeed, the AB signals to the hippocampus the experiences that are more emotionally salient and should have a sort of priority for storage, relative to other poorly salient experiences that are less significant for the individual. As a consequence, the AB is thought to be involved in the genesis of the affective symptomatology of many neuropsychiatric disorders, such as trauma spectrum and mood spectrum disorders, in which the emotional features of mnesic functions are often altered [14–17]. Accordingly, the AB and the connections of its functional network have been successfully used as a target for Deep Brain Stimulation (DBS) in refractory cases of some of these conditions [18–26].

In this scenario, gaining a deeper understanding of the morphophysiology of the AB has become increasingly relevant not only for neuroscience researchers but also for the clinicians who, in their daily practice, deal with the aberrations of such fascinating aspects of neurobiology.

2. Neuroanatomy

The AB is located deep within the anterior segment of the uncus and it appears as an ovoid-shaped gray matter nuclear complex with an anteroposterior and transverse diameter of 12 mm and of 16 mm, respectively. The medial surface of the AB is in relation to the semilunar and ambiens gyri. Laterally, the posterior aspect of the AB faces the intralimbic gyrus and piriform cortex, (**Figure 2a**) and medially constitutes the anterolateral wall of the temporal horn of the lateral ventricle. This surface is concave due to the presence of the head of the hippocampus, from which it is separated by the uncal recess (**Figure 2b**). In the temporal horn, the superior portion of the AB is within the anterior roof of the ventricle, and anteriorly within the white matter of the temporal pole (**Figure 2b**). Superiorly, the AB is continuous with the globus pallidus and with the ventral claustrum (**Figure 3**) [27].

2.1 Microneuroanatomy

Burdach was the first, in the early nineteenth century, to provide a comprehensive description of this structure, for which he proposed the name "amygdala" in view of its resemblance to an almond seed [28–31]. During the last century, several authors have improved the anatomical knowledge of the region, identifying a large number of heterogeneous nuclei surrounding the one originally described by Burdach. Accordingly, the name of this area of gray matter was changed to the more appropriate term AB [31].

This nuclear complex is currently subdivided into four groups of nuclei: a basolateral complex, which is the one originally described by Burdach, a centromedial complex, a cortical complex, and an additional fourth group [27, 31], as shown in **Figure 3**.

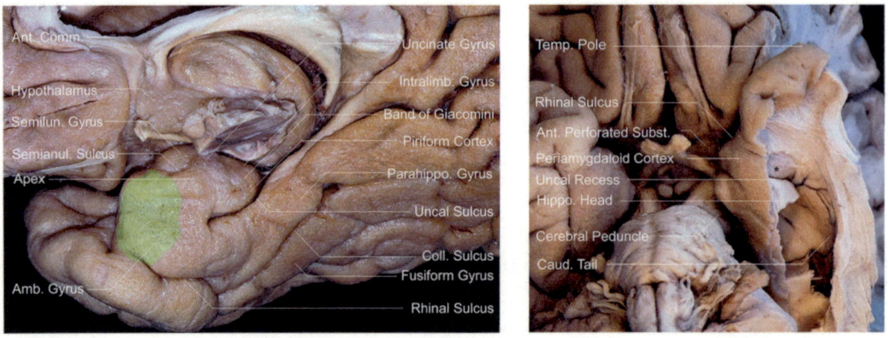

(a) (b)

Figure 2.
Morphology and location of the amygdaloid body (*). In "a", exposure of the medial temporal region is shown. The amygdaloid body is located deep within the anterior segment of the uncus. The relation between the amygdaloid body and the periamygdaloid cortex is highlighted in green. The medial surface of the amygdaloid body is adjacent to the semilunar gyrus, superiorly, and the ambiens gyrus, inferiorly. Postero-medially and postero-inferiorly, the amygdaloid body faces the intralimbic gyrus and the piriform cortex, respectively. In "b", the dissection of the basal aspect of the left temporal lobe is shown. The relation between the amygdaloid body and the surrounding structures is highlighted. The uncus consists of an anterior portion, containing the amygdaloid body and covered by the peri-amygdaloid cortex, and a posterior portion, enclosing the head of the hippocampus and covered by the cortex of the parahippocampal gyrus. The amygdaloid body is separated from the head of the hippocampus by the uncal recess. Anteromedially, the amygdaloid body is separated from the anterior perforated substance by the rhinal sulcus.

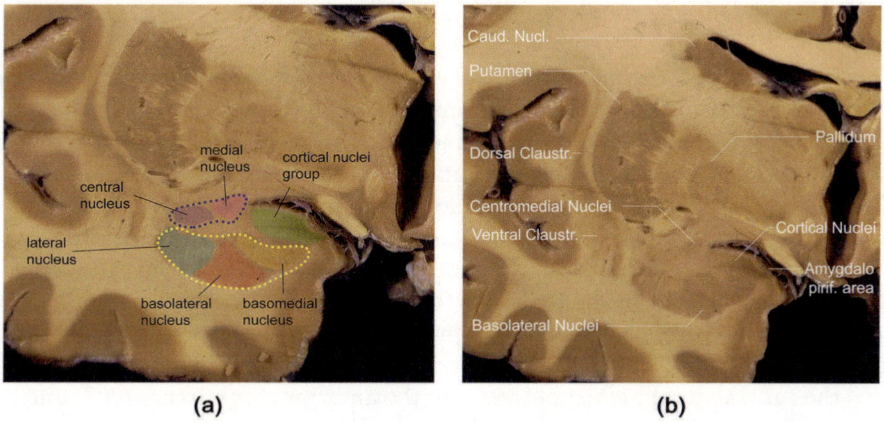

Figure 3.
Coronal cut of the medial temporal lobe at the level of the anterior commissure showing the nuclear subdivision of the amygdaloid body, in macroscopic direct-view. In "a", the amygdaloid nuclei are highlighted. The basolateral amygdaloid body (yellow dotted line) is composed of the lateral (blue), basolateral (red), and basomedial (yellow) nuclei, disposed in a latero-medial direction. The lateral nucleus exhibits a striate appearance, due to the passage of fibers of the ventromedial portion of the uncinate fascicle. The centromedial amygdaloid body (blue dotted line) is composed of the central nucleus (purple), located infero-laterally and anteriorly, and the medial nucleus (pink), located superomedially and posteriorly. The cortical amygdaloid body (green) comprises multiple tiny nuclei located superomedially, in the context of the nuclear complex. In "b", the relation between the amygdaloid body and the surrounding structure is highlighted. The ventral claustrum is scattered throughout the uncinate fascicle, without a clear demarcation from the basolateral amygdaloid body. The amygdalo-piriform area is composed of the gray matter located between the amygdaloid body and the piriform cortex. Superiorly, the amygdaloid body adjoins the white matter of the basal forebrain and it is continuous cranially with the globus pallidus.

The basolateral AB has a telencephalic origin and is composed by the lateral, basolateral, and basomedial nuclei. The lateral nucleus exhibits a striate appearance, due to the presence of fibers of the ventromedial uncinate fascicle.

The centromedial AB has a diencephalic origin and is composed of the central nucleus, located inferolaterally and anteriorly, and the medial nucleus, located superomedially and posteriorly, adjacent to the stria terminalis. Both nuclei are pale, reminding the color of the pallidum with which they share the ontogenetic origin.

The cortical AB has a telencephalic origin and is composed by multiple tiny nuclei located superomedially, in the context of the nuclear complex. These nuclei are continuous with the piriform cortex of the uncus, and include the primary and secondary nuclei of the lateral olfactory stria and the anterior cortical amygdaloid nucleus intercalated in the olfactory system.

The additional group comprises all gray matter areas surrounding the proper nuclear complex, and seamlessly transitions into the neighboring structures. It includes two essentially independent structures: the telencephalic amygdalo-piriform area, located between the AB and the piriform cortex; and the diencephalic bed nucleus of the stria terminalis, which identifies with the gray matter following the course of the stria terminalis and constitutes the continuity between the AB and the hypothalamus.

2.2 Connectomic

The AB is characterized by numerous connections with the limbic system, the thalamus, the brainstem, and the neocortex, organized in a three-dimensional chassis around the nuclear complex [12, 30].

2.2.1 The amygdalofugal pathways

Humans and other mammals are endowed with two divergent amygdaloid projection systems named dorsal and ventral amygdalofugal pathways [30–33]. These two bundles arise from the centromedial AB and from the basolateral AB, respectively, and converge at the level of the septum and of the hypothalamus [30].

The dorsal amygdalofugal pathway, shown in **Figure 4**, forms the stria terminalis. This projection system has been comprehensively described in humans [27, 30, 31] as a C-shaped bundle arising from the centromedial AB and following the lateral ventricle and the surface of the thalamus toward the ipsilateral foramen of Monro. Hence, it splits into three components: a pre-commissural component, reaching the septum and the nucleus accumbens septi; a commissural component, running toward the contralateral hemisphere within the anterior commissure (AC) and taking part in the inter-amygdaloid pathway; and a post-commissural component running behind the AC. The latter, splits into a portion following the post-commissural fornix to reach the mammillary body, and another portion following the stria medullaris thalami to reach the habenula, the thalamus, and the hypothalamus. All along its course, the stria terminalis is in continuity with its bed nucleus. The bed nucleus of the stria terminalis is subdivided into a rostral, intermediate, and caudal portions and it represents the anatomical gray matter continuity between the AB and the hypothalamus.

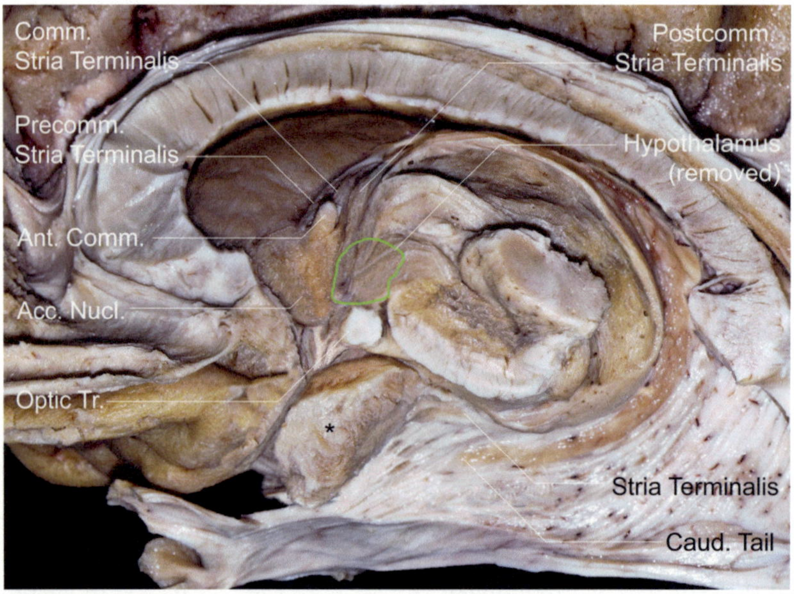

Figure 4.
Dissection of the medial aspect of the right hemisphere. The complete course of the stria terminalis is exposed after the complete removal of the fornix and the hypothalamus (which position is still marked in green). The stria terminalis is a C-shaped bundle arising from the centromedial amygdaloid body (*) and following the lateral ventricle and the surface of the thalamus toward the ipsilateral foramen of Monro, where it splits into three components: A pre-commissural component, reaching the septum and the nucleus accumbens septi; a commissural component, running toward the contralateral hemisphere within the anterior commissure and taking part of the inter-amygdaloid pathway; and a post-commissural component running behind the anterior commissure. The latter splits in a portion following the post-commissural fornix and another portion following the stria medullaris thalami to reach the mammillary body, the habenula, the thalamus, and the hypothalamus.

The ventral amygdalofugal pathway arises from basolateral AB and runs ventrally to the ventral striato-pallidal structures, the AC, and the internal capsule, to reach the septal region. Due to the spread diffusion of its fibers into several structures and due the limited size of the region, a comprehensive description of its anatomy has always been difficult to achieve [27, 30, 31]. Using Klinger's fiber dissection technique on ex-vivo brain specimens, we have recently provided a new account on the subject [27]. According to our findings, the ventral amygdalofugal pathway appears as a widespread bundle connecting the basolateral AB with the septum, the thalamus and the hypothalamus, characterized by two arms diverging below the AC, as shown in **Figure 5**.

The anterior arm of the ventral amygdalofugal pathway, named also diagonal band of Broca, runs underneath the AC in a latero-medial direction toward the basal nucleus of Meynert (scattered along its course) and bends anteriorly and medially toward the nucleus accumbens septi, directed to the septum. The posterior arm of the ventral amygdalofugal pathway runs laterally above the anterior perforated substance and toward the thalamus and hypothalamus, where it intermingles with fibers of the medial forebrain bundle (MFB) and of the stria medullaris thalami.

2.2.2 Connections with the brainstem

The MFB connects the AB with the brainstem [4, 34], as shown in **Figure 5**, and is supposed to play a role in the neurovegetative correlates of implicit memories. Since the rostral portion of the MFB splits into several different branches, extending in various directions and depths, its dissection in humans has always been partial [30, 35–38]. In the diffusion MRI tractographic study performed by Coenen et al. [39] the MFB appeared composed by a main trunk arising in the dentate nucleus of the cerebellum, and running through the retrobulbar area and the periaqueductal gray, to reach the ventral-tegmental area of the midbrain. Here, it splits into an inferomedial portion, running along the lateral wall of the third ventricle and reaching

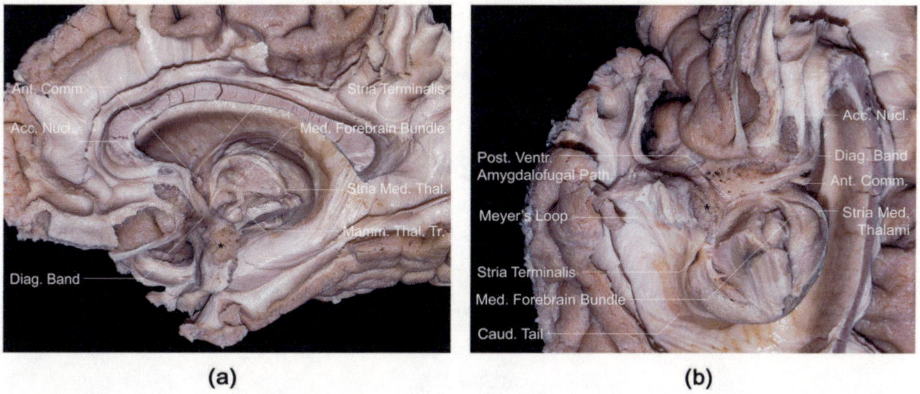

Figure 5.
In "a" and in "b", the dissection of the medial and inferior aspects of the same right hemisphere are highlighted, respectively. Both pictures show the complete course of the ventral amygdalofugal pathway, which arises from the basolateral amygdaloid body (*) and immediately spreads into two arms. The anterior arm of the ventral amygdalofugal pathway, named also diagonal band of Broca, runs underneath the anterior commissure in a latero-medial direction toward the basal nucleus of Meynert (scattered along its course) and bends anteriorly and medially toward the nucleus accumbens septi, directed to the septum. The posterior arm of the ventral amygdalofugal pathway runs laterally above the anterior perforated substance and toward the thalamus and hypothalamus, where it intermingles with fibers of the medial forebrain bundle and of the stria medullaris thalami.

the hypothalamus; and in a superolateral portion, running inferolaterally to the thalamus. In particular, the superolateral MFB has been dissected in ex-vivo specimens, from the ventral-tegmental area, caudally, to the medial ceiling of the anterior perforated substance, rostrally, where it intermingles with the posterior portion of the ventral amygdalofugal pathway [27]. Its neuromodulation in humans seems to show mood-regulating effects in refractory major depressive disorder (MDD) [25].

2.2.3 Connections with the hippocampus

The intra-ventricular amygdalo-hippocampal bundle, shown in **Figure 6**, connects the basolateral AB with the head of the hippocampus. This connection was firstly described in rhesus monkeys by Rosene and Van Hoesen in 1977 [40], and in rats by Kemppanien et al. in 2002 [41]. In humans, it was identified by Di Marino et al. in 2016 [31], and by our group in 2021 [27]. Kemppainen and Pitkanen [42] demonstrated that the connections between the basolateral AB and the CA1/subiculum of the hippocampus are resistant to neuronal damage induced by the status epilepticus in rats, suggesting their role as a pathway for seizure propagation between the two structures. Di Marino et al. [31] suggested the existence of an additional extra-ventricular amygdalo-hippocampal connection, whose existence has still to be confirmed [27, 31].

2.2.4 Cortical connections

The amygdalo-temporal fascicle, whose function is still unknown, arises from the basolateral AB and runs straight to the cortex of the temporal pole, as shown in **Figure 6**. Its existence in humans was first mentioned by Curran in 1909 [43], but it has been described in detail only by Klingler and Gloor in 1960 [30] and by our group in 2021 [27].

The temporo-pulvinar bundle of Arnold arises from the anterior temporal cortex with a postero-medial course and is arranged in a thin sheet of fibers running above the roof of the temporal horn underneath the tail of the caudate nucleus, establishing a connection with the lateral AB. Afterwards, it continues posteriorly within the

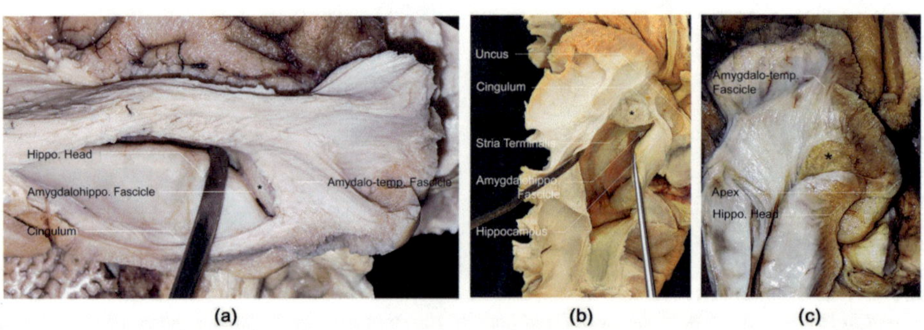

Figure 6.
In "a", the lateral dissection of the right hemisphere is shown, while in "b" and in "c", its stepwise basal dissection is shown. In "a", the removal of the inferior longitudinal fascicle allows the exposure of the amygdaloid body (*), anteriorly and of the hippocampus posteriorly. Retracting the hippocampus postero-inferiorly with a spatula was possible to expose the tiny bundle that connects the aforementioned structures. Such connection is exposed also in "b", where the cingulum has been partially sectioned and the hippocampus retracted medially (exposing the roof of the temporal horn of the lateral ventricle). "c" Highlights the amygdalo-temporal fascicle.

sublenticular segment of the internal capsule, eventually ending into the pulvinar. Its dissection is shown in **Figure 7**. Such bundle was first detected by von Monakow in 1895 [44] and extensively described by Klingler and Gloor [30], who termed it "inferior thalamic peduncle," and by our group [27]. Its function is still unknown, but clinical evidences of neuromodulation suggest that it could play a role in mood control [45].

The lateral olfactory stria, whose dissection is shown in **Figure 7a**, arises from the olfactory tract, in front of the anterior perforated substance, and represents the lateral part of the olfactory trigon. It runs laterally, just anteriorly to the diagonal band, to intermingle with fibers of the external capsule at the level of the limen insulae and insular gyri. Laterally, the lateral olfactory stria provides fibers directed toward the cortical AB, in particular, it reaches the primary and secondary olfactory nuclei and the anterior cortical amygdaloid nucleus. Along its course, the lateral olfactory stria interconnects the olfactory bulbs, the gray matter of the anterior perforated substance (considered the involution of the laminated-olfactory tubercle of macrosmatic mammals), the piriform cortex, the AB, and the insula; for this reason, Nieuwenhuys et al. [4] have considered such pathway as the anatomical substrate of the link between taste and smell, as well as between smell, emotions, and memories. It is widely ascertained that olfactory clues are the strongest inputs for perceptual recalling of emotional components of memories (commonly referred to by the French as Proust phenomenon) [46]. In fact, the lateral olfactory stria connects the olfactory bulb directly with the AB and the piriform cortex, making olfaction the sole sensory modality to reach the cortex without thalamic retransmission.

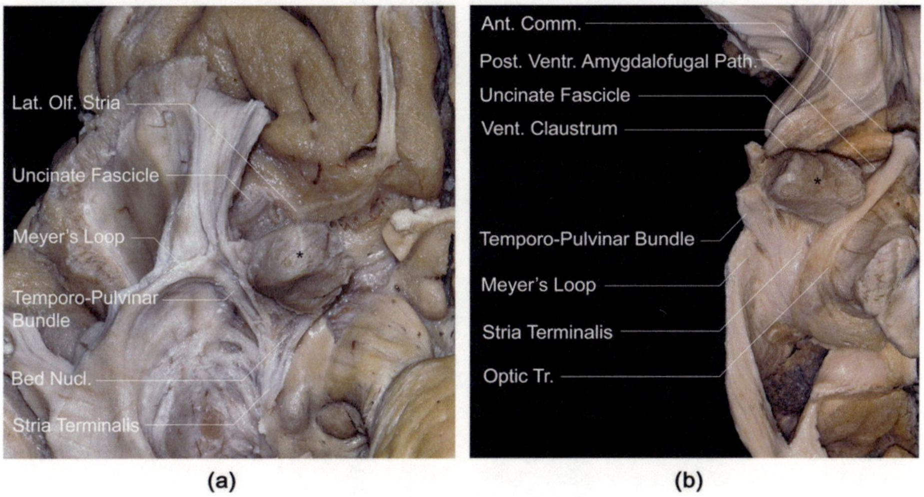

Figure 7.
Dissection of the inferior aspect of the right hemisphere. "a" Highlights the relation between the amygdaloid body (*) and the uncinate fascicle, located anterolaterally, the lateral olfactory stria, located anteromedially, and the temporo-pulvinar bundle of Arnold located postero-laterally. The lateral olfactory stria arises from the olfactory tract, in front of the anterior perforated substance and runs laterally, just anteriorly to the diagonal band, to intermingle with fibers of the external capsule at the level of the limen insulae and insular gyri. Laterally, the lateral olfactory stria provides fibers directed toward the cortical amygdaloid body. The temporo-pulvinar bundle of Arnold arises from the anterior temporal cortex with a postero-medial course and is arranged in a thin sheet of fibers running above the roof of the temporal horn underneath the tail of the caudate nucleus, establishing a connection with the lateral amygdaloid body. Afterwards, it continues posteriorly within the sublenticular segment of the internal capsule, eventually ending into the pulvinar. In "b", the lateral olfactory stria has been sectioned exposing the anterior commissure.

Herrick in 1956 wrote: "This olfactory field at the anterior end of the brain is the dominant centre of control of all behaviours of these primitive vertebrates, and for this reason it was the seedbed for further structural differentiation as the patterns of behaviour were stepped up from one integrative level to another. Here, the rudimentary cortex had its beginnings" [5].

The uncinate fascicle (UF), whose dissection is shown in **Figure 7** and in **Figure 8**, connects the frontal lobe with the anterior portion of the temporal lobe. It is composed by a ventromedial part, arising from the medial orbitofrontal cortex (mOFC), and an anterolateral part, arising from the medial prefrontal cortex (mPFC). In its course to the temporal pole, the UF bends below the nucleus accumbens septi forming the supero-anterior wall of the Gratiolet's canal, where some fibers of its ventromedial part provide connections with the basolateral AB, as shown in **Figure 8c**. The ventral claustrum is scattered throughout the ventromedial UF, without a clear demarcation from the AB. The fibers of the anterolateral UF follow their course straight to the cortex of the temporal pole [27, 30].

The most conspicuous AB's cortical connection is constituted by the cingulum, whose dissection is shown in **Figure 9**. The cingulum, in humans, arises from the basolateral AB and runs over the medial surface of the hippocampus in a postero-medial direction (as shown in **Figure 9b**) interconnecting the gyrus ambiens and the parahippocampal gyrus. Afterwards, it bends anteriorly, above the corpus callosum, reaching first the parietal cortex, then the mPFC, eventually ending at the level of

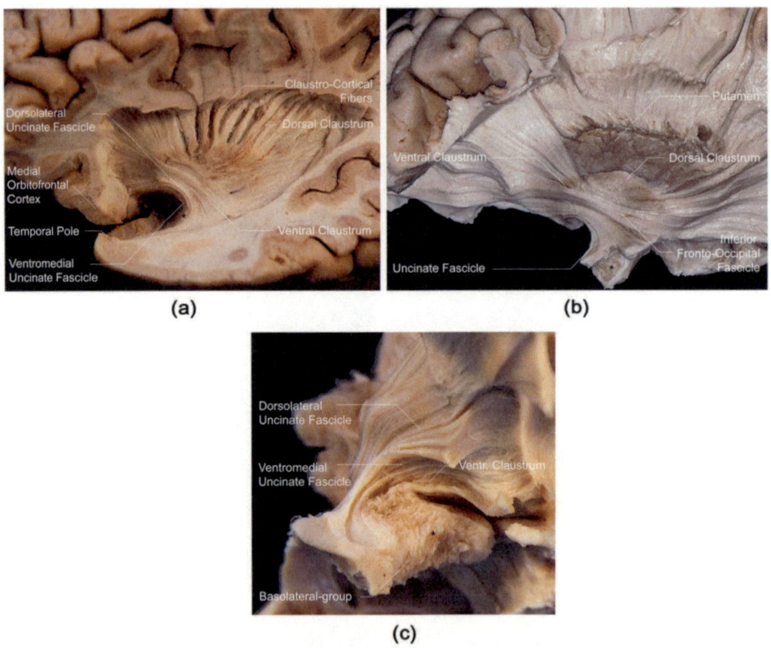

Figure 8.
The dissection of the lateral aspect of the left hemisphere. In "a", the whole course of the uncinate fascicle that connects the frontal and the temporal lobes is shown. Its relation with the ventral claustrum, scattered throughout its course without a clear demarcation from the amygdaloid body (*), is highlighted in "b". The uncinate fascicle is composed by a ventromedial part, arising from the medial orbitofrontal cortex, and an anterolateral part, arising from the medial prefrontal cortex. In its course to the temporal pole, it bends below the nucleus accumbens septi forming the supero-anterior wall of the Gratiolet's canal, and, at this point, some fibers of its ventromedial part provide connections with the basolateral amygdaloid body as shown in "c".

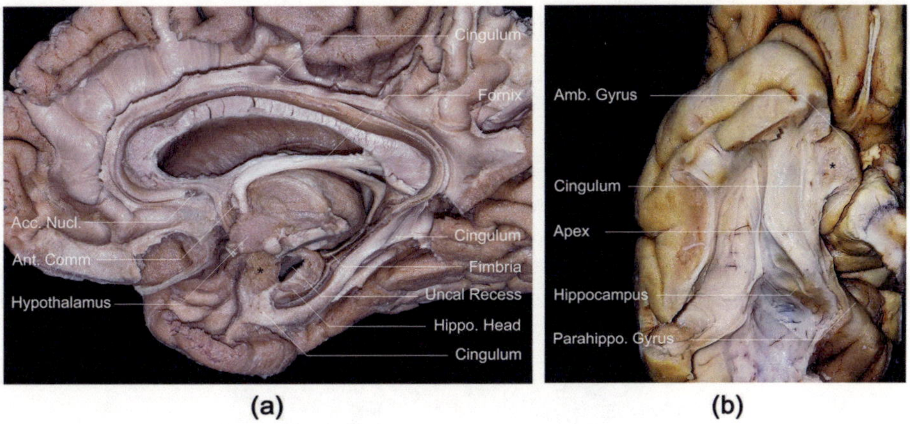

Figure 9.
The dissection of the medial aspect of the right hemisphere. In "a", the whole course of the cingulum is shown, which arises from the basolateral amygdaloid body (*), and runs over the medial surface of the hippocampus in a postero-medial direction interconnecting the gyrus ambiens and the parahippocampal gyrus. Afterwards, it bends anteriorly, above the corpus callosum, reaching first the parietal cortex, then the medial prefrontal cortex, eventually ending at the level of the basal forebrain. The arising point of the cingulum from the basolateral surface of the amygdaloid body is shown in "b".

the basal forebrain [27]. Since the observation of Dejerine in 1895, there has been the idea that the cingulum would be composed by the combination of short fibers, arising from neighboring gyri [32, 47]. This observation was confirmed in 2014 by the tract-tracing study performed by Heilbronner and Haber [48], and in 2021 by our dissection work [27], which illustrated fibers arising from the basolateral AB getting directly into the temporal cingulum, and indirectly, through the ventral amygdalofugal pathway, into the rostral-subgenual cingulum on the medial surface of the nucleus accumbens septi.

The ABs are reciprocally connected by the AC, the major inter-hemispheric connection of the limbic system. The AC, which course is shown in the dissection in **Figure 10**, is a transverse, handlebar-shaped bundle composed by a compacted medial portion that spreads bilaterally into the temporal lobes. The medial AC is located in front of the columns of the fornix, at the rostral edge of the third ventricle, and is housed into a canal formed entirely by white matter, named canal of Gratiolet. Such canal is composed by the UF, anteriorly and superiorly, from the diagonal band of Broca anteriorly and inferiorly, and from the columns of the fornix, posteriorly. From its medial portion, the AC curves bilaterally and posteriorly passing under the lenticular nucleus before crossing the temporal isthmus. At this point, it splits into an anterior crus extending forword to the mPFC, and in a fan-shaped posterior crus giving branches getting into the AB's dorsal surface [27, 30, 49]. Eventually, the posterior crus divides into an occipital extension, joining the sagittal stratum, and a temporal extension, reaching the anterolateral-temporal cortex.

2.3 Functional anatomy of the AB

The heterogeneous pattern of projections of the AB has been recently organized in two different functional systems by de Olmos and Heimer [50]: the extended amygdala and the great limbic lobe.

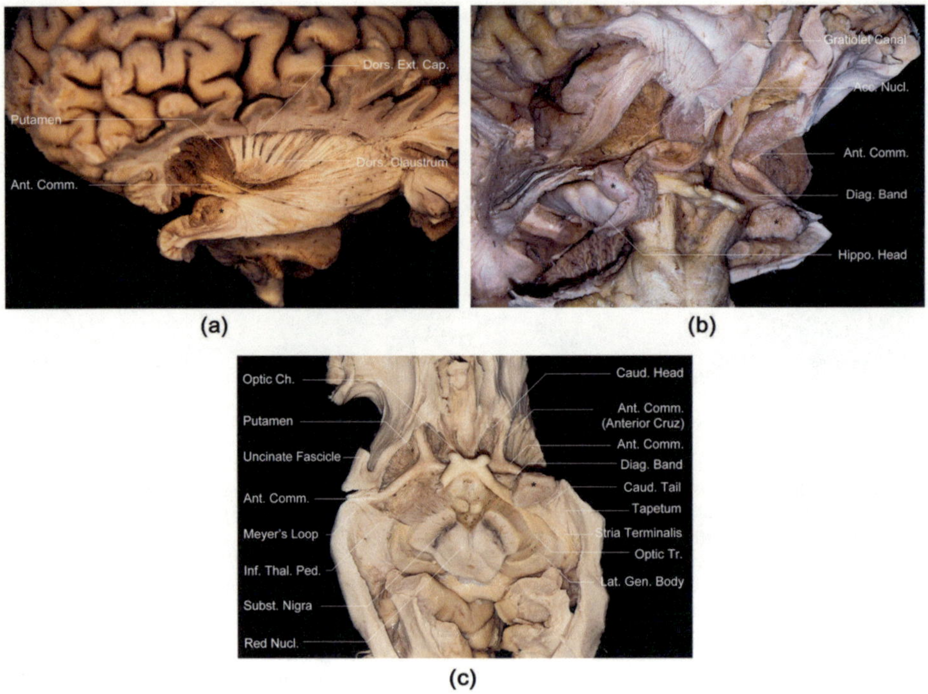

Figure 10.
The dissection of the lateral aspect of the left hemisphere in "a" and of the inferior aspect of the brain in "b" and "c" show the complete course of the anterior commissure starting from its compacted medial part housed into the Gratiolet's canal and spreading bilaterally into and anterior and a posterior course reaching the frontal and temporal lobes and interconnecting the two amygdaloid body.

The extended amygdala belongs to the ventral pallidum and is constituted by a continuum of gray matter of diencephalic origin organized as a ring encircling the internal capsule. It is composed by the centromedial AB, the bed nucleus of the stria terminalis, and the lateral and medial hypothalamus. All structures belonging to the extended amygdala are shown in the dissection in **Figure 11**. These structures are reciprocally connected through the stria terminalis, which is supposed to link food-related, olfactory, sexual, and endocrine cues [12, 50]. In particular, the central AB seems to be directly connected to the autonomic and somatomotor centers in the lateral hypothalamus, while the medial AB seems to be directly connected to the endocrine-related medial hypothalamus.

The great limbic lobe, shown in **Figure 12**, includes the basolateral AB, the cortical AB, the hippocampus, the basal forebrain, the insula, the cingulate gyrus, the ventral prefrontal cortices, the subcallosal, and the entorhinal cortices. All these structures have a telencephalic origin, and its subcortical nuclei belong to the ventral striatum. The white matter bundles connecting the structures of the great limbic lobe are the ventral amygdalofugal pathway, the cingulum, the UF, the MFB, and the lateral olfactory stria [12, 50].

The converging point between the extended amygdala and the great limbic lobe is the basolateral AB, that receives afferent fibers from the cortex and from the ventral striatum, and projects efferent fibers to the centromedial AB.

In view of the recent advances in functional neurosurgery, such classification acquires a clinical relevance. Indeed, the basolateral AB has been used as DBS-target

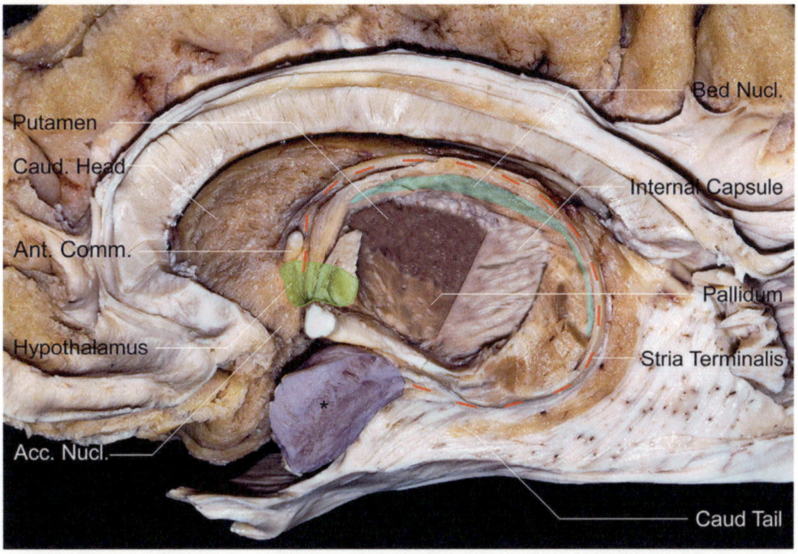

Figure 11.
Shows the structures composing the extended amygdala that belongs to the ventral pallidum and is constituted by a continuum of gray matter of diencephalic origin organized as a ring encircling the internal capsule. It is composed by the centromedial amygdaloid body (*), the bed nucleus of the stria terminalis (light blue) and the lateral and medial hypothalamus (green). These structures are reciprocally connected through the stria terminalis (red dashed line).

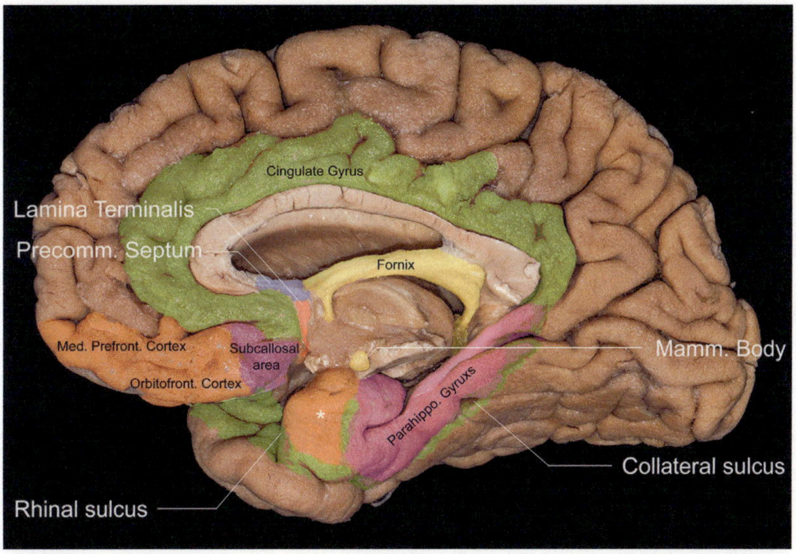

Figure 12.
Shows the brain cortical regions composing the great limbic lobe that includes the basolateral and cortical amygdaloid body (*), the hippocampus, the basal forebrain, the insula, the cingulate gyrus, the prefrontal cortices, and the subcallosal, entorhinal, and ventral-tegmental areas.

for the treatment of refractory Post-Traumatic Stress Disorder (PTSD), [22, 23, 26] autism, [19–21] and, together with the cingulum and the superolateral portion of the MFB, refractory MDD [18–20, 51]. Furthermore, DBS targeting the bed nucleus

Paper	Year	Target	Disorder	Effect
Kennedy et al. [18]	2011	Basolateral amygdaloid body	Autism	Improve
Langevin et al. [19]	2012	Basolateral amygdaloid body	Autism	Improve
Langevin et al. [19]	2012	Cingulum	MDD	Improve
Schlaepfer et al. [20]	2013	Basolateral amygdaloid body	Autism	Improve
Schlaepfer et al. [20]	2013	Basolateral amygdaloid body	MDD	Improve
Sturm et al. [21]	2013	Superolat. Portion of the MFB	MDD	Improve
Koek et al. [22]	2014	Basolateral amygdaloid body	PTSD	Improve
Langevin et al. [23]	2015	Basolateral amygdaloid body	PTSD	Improve
Blomstedt et al. [25]	2017	Medial forebrain bundle	MDD	Improve
Lavano et al. [26]	2018	Basolateral amygdaloid body	PTSD	Improve
Piacentiani et al. [53]	2008	Stria terminalis (unintentional)	Dystonia	Developing of MDD with psychosis
Luyten et al. [24]	2016	Bed nucleus of the stria terminalis	DOC	Improve
Blomstedt et al. [25]	2017	Bed nucleus of the stria terminalis	Relapsing anorexia	Improve

Table 1.
Summary of the clinical outcomes in patients treated with the DBS of the amygdaloid body and its connections.

of the stria terminalis has been shown effective in decreasing anxiety in patients with obsessive-compulsive disorder (DOC) [24] and relapsing anorexia [25]. These observations are in line with the results of analogous studies in animals [52]. On the other hand, Piacentiani et al. [53] reported that the unintentional stimulation of the stria terminalis, caused by the displacement of a DBS-electrode previously positioned in a patient affected by dystonia, determined the development of MDD with mood-incongruent psychotic symptoms. These pieces of evidence support the idea that the stimulation of the AB's connections, comprised in the great limbic lobe, might increase mood control and decrease the severity of aversive emotional responses to traumatic memories [20–23, 25, 26, 51, 54–57]. On the contrary, the effects of the stimulation of the extended amygdala in humans are still controversial and the few case reports in the literature are not enough to predict its possible outcomes [24, 25, 53]. All results are resumed in **Table 1**.

3. Role of the AB in trauma spectrum disorders

Trauma spectrum disorders are a group of psychological alterations consequent to the exposure to a trauma, defined as an event that is perceived to be life-threatening or to pose the potential of serious bodily injury to self or others [58]. This condition is considered as a chronic dysregulation of the physiologic nervous response that normally follows the exposure to a threat.

Trauma spectrum symptomatology is characterized by phenomena that can be grouped into three primary domains: *reminders of the exposure* such as flashbacks, intrusive thoughts, and nightmares; *activation patterns* such as hyperarousal, insomnia, agitation, irritability, impulsivity, and anger; and *deactivation patterns* such as numbing, avoidance, withdrawal, confusion, derealization, depression, and dissociation. In particular, dissociation is defined as a failure, disruption, interruption, and/or discontinuity of the normal, subjective integration of behavior, memory, identity, consciousness, emotion, perception, body representation, and motor control, and it represents the link between trauma spectrum disorders [58] and a large variety of psychiatric disorders like drug abuse, self-injurious behaviors, suicidality, somatization, and borderline personality disorder [59, 60].

Dissociative disorders can be classified into two subtypes: structural dissociation, involving the autonomic responses following the exposure of traumatic cues, leading to possible repercussion on personality and general functioning; and somatoform dissociation, which is characterized by the expression of implicit viscerosomatic memories, leading to body representation disorders and to a large variety of dermatologic and immune disorders [59, 60].

Clinical manifestation of the psychological trauma has been known since the beginning of the psychoanalysis [2], although the deep insight of its consequences on the society was reached only in the second part of the nineteenth century when it has been given a bigger attention to war veterans difficulties at the moment of their return to daily living. The huge increase of the post-traumatic diagnoses in the last decades, whose recognition has moved far beyond the border of war situations, along with their heavy potential for social impairment, highlighted the need for a more accurate nosology and a deeper insight into the pathophysiology of these disorders.

It is currently thought that there is a substantial continuity in the expression of trauma spectrum disorders and, accordingly, trauma manifestations have been organized in two groups of clinical severity: the simple and the complex PTSDs [58–60].

Simple PTSDs are characterized by the presence of a single or multiple traumatic event/s that occurred after the psychological maturity of the subject. These individuals typically retain both explicit and implicit memories of the trauma. Simple PTSDs comprise the acute post-traumatic disorder and the proper PTSD.

Complex PTSDs are characterized by the presence of a single or more often multiple *relational* traumatic event/s that occurred before the psychological maturity of the subject. Individuals with Complex PTSD typically retain implicit memories of the traumatic event(s), but not necessarily explicit memories. Complex PTDSs comprise the proper complex PTSD, the Dissociative Amnesias, the depersonalization/derealization disorders, and the dissociative identity disorder.

This classification has been supported by functional magnetic resonance studies, which have observed differences in the cerebral activation pattern between the simple and complex PTSDs [59–61]. According to these studies, when exposed to traumatic cues, a subject affected by a simple PTSD presents a hyperactivation of the AB, a relative hyperactivation of the autonomic nervous system (ANS), and a deactivation of the mPFC; differently, in a subject affected by a complex PTSD, the activation pattern might not be so well defined and a hyperactivation of the AB/ANS might not be so evident, the thalamus might be deactivated, and some parts of the cortex might present a different degree of activation/deactivation.

These neurobiological evidences found clinical confirmation in the symptomatic expression of the two groups of post-traumatic disorders: in simple PTSDs, the symptoms consequent to the reminder of the exposure are prevalent, while in complex PTSDs the symptomatic pattern is more characterized by hyperactivation and deactivation [58, 62].

In **Table 2** the biological evidences reported in the PTDS are resumed. The endocrine system presents activation of the hypothalamic-pituitary-adrenal axis and of the hypothalamic-pituitary-thyroid axis. The cerebral neurochemical pattern is characterized by increased levels of noradrenalin and by decreased levels of serotonin. Eventually, some parts of the brain like the hippocampus, the mPFC, and the anterior cingulate gyrus present a volume reduction [62].

Feature	Change	Effect
1. Neuroendocrine		
Hypothalamic-pituitary-adrenal axis	Hypocortisolism	Disinhibits CRH/NE and upregulates response to stress Drives abnormal stress encoding and fear processing
	Sustained, increased level of CRH	Blunts ACTH response to CRH stimulation Promotes hippocampal atrophy
Hypothalamic-pituitary-thyroid axis	Abnormal T3:T4 ratio	Increases subjective anxiety
2. Neurochemical		
Dopamine	Increased levels	Interferes with fear conditioning by mesolimbic system
Norepinephrine	Increased levels	Increases arousal, startle response, encoding of fear memories
Serotonin	Decreased levels in dorsal and median raphe	Disturbs dynamic between amygdala and hippocampus Compromises anxiolytic effects Increases vigilance, startle, impulsivity, and memory intrusions
GABA	Decreased levels	Compromises anxiolytic effects
Glutammate	Increased levels	Derealization and dissociation
NPY	Decreased levels	Leaves CRH/NE unopposed and upregulates response to stress
β-endorphin	Increased levels	Numbing, stress-induced analgesia, and dissociation
3. Neuroanatomical		
Hippocampus	Reduced volume and activity	Alters stress responses and extinction
Amygdaloid body	Increased activity	Promotes hypervigilance and impairs discrimination of threat
Prefrontal cortex	Reduced volume and activation	Dysregulates executive functions
Anterior cingulate cortex	Reduced volume	Impairs the extinction of fear responses

Table 2.
Biological evidences reported in patients affected by PTDS.

This piece of knowledge has paved the way for the development of numerous theories regarding the pathophysiology of each AB's connection, opening up opportunities for their clinical experimental use as DBS-targets in neuromodulation [23, 27, 62, 63].

In particular, the UF and the cingulum seem involved in the process that links implicit and explicit memories. A conduction disorder of these bundles seems to be involved in the genesis of the symptoms consequent to the reminders of the exposure and their neuromodulation showed promising results in the treatment of refractory-PTSD and MDD [19, 59].

The connections between the basolateral AB and the thalamus, such as the temporo-pulvinar bundle of Arnold and, especially, the ventral amygdalofugal pathway, seem to play a role in thalamic deactivation, and, consequently, in the development of the symptomatic deactivation pattern of PTSD [59, 61, 64].

The MFB, connecting the basolateral AB directly with the brainstem, and the stria terminalis, connecting the centromedial AB with the lateral hypothalamus could constitute a direct and an indirect links between the exposure of traumatic cues and the activation of the ANS, whose nuclei are located in the brainstem. Some authors [6, 60, 64] supported the idea that these pathways could be involved in the genesis of structural dissociation.

Additionally, the portion of the stria terminalis, connecting the centromedial AB with the medial hypothalamus, seems to be involved also in the post-traumatic endocrine dysregulation and, together with the temporo-insular connections (involved in visceral introjection belonging to the vagus nerve) could play a role in the development of the somatoform dissociation [65].

4. Role of the AB in mood disorders

The neuroanatomical and neurofunctional studies related to mood disorders are much compounded by the Leonhardian dichotomy, adopted by the DSM tradition, which categorically distinguishes between bipolar and unipolar mood disorders. As a matter of fact, the literature about the structural and functional brain correlates of both nosological groups does not yield significantly different results [66, 67] although recent imaging studies claim to have found subtle pattern differentiations [68–70]. One of the most consistent findings, when we examine the role of the amygdaloid body in mood disorders, is the tendency toward an overactivity and progressive atrophy of this brain region, particularly on the left side [66, 67, 71], but these alterations are to be mindfully considered in light of the astounding heterogeneity that characterizes mood disorders. It is possible that these amygdalar dysfunctions are not related to a specific disease, being in fact transcategorial biomarkers of dysphoric experiences, such as hyperarousal, anxiety, and irritability [71]. It is also presumable that the amygdaloid body could be more implicated with some subsets/endophenotypes of mood disorders, while being utterly marginal in other pathotypes [71]. Finally, it is likely that some changes might be state-specific (mania, hypomania, mixed states, depression), whereas other changes may be trait-related and, therefore, underlying all mood states [72]. Subsequently, while waiting for more neurobiologically-based psychopathological dimensions, addressing amygdala's implications in affective disorders as a whole heterogeneous group seems to be the most convenient approach.

The first sound experimental contributions concerning the role of the amygdaloid body in manic-depressive illness are to be ascribed to the research of Post and coworkers [73, 74]. Largely drawing from observations of progressive epileptic threshold lowering in murine models of temporal lobe epilepsy, these authors prompted that overexcitability in the amygdaloid region could be engendered by long-term repetition of subthreshold stimuli (e.g. minor life events or subclinical mood states), at least in subjects with a substantial degree of constitutional (familial) predisposition. According to this perspective, amygdala-kindled seizures would represent a non-homologous model for mood disorders recurrence and mood episodes could be heuristically interpreted as "prolonged non-convulsive limbic pseudo-epilepsies." Post's theoretical model is partly prescient of later acquisitions and was historically crucial for the consolidation of the concept of mood-stabilization with antiepileptic drugs [75, 76]. A more thorough neuroanatomical model of BD was proposed by Stephen Strakowski et al. in 2012 [77]. According to their synthesis, mostly based on brain imaging studies, amygdala's dysregulation could be placed at the center of two major emotion-regulatory cortico-striato-pallido-thalamo-cortical networks: the ventrolateral prefrontal pathway, linked to explicit and conscious emotional regulation, and the orbitofrontal pathway, associated with implicit and non-conscious affective processing [77]. This model describes essentially a top-down/bottom-up imbalance hinged upon fronto-limbic structures: a neurodevelopmental fragility in top-down regulation of the amygdaloid body and related structures, caused by an ineffective maturation of ventral prefrontal cortices or related white matter systems, would predispose subjects to extreme affective states, wherein subcortical limbic centers are insufficiently modulated and are thus free to provoke a sort of bottom-up, emotion-laden neural flooding.

Morphometric studies report contradictory findings at a first glance, showing either reductions or increases in amygdalar volumes among subjects with BD [67]. These results are to be understood with some cautiousness, since the adults with BD recruited in these studies might conceal a host of confounding factors, including differences in disease cyclicity and polarity or the differential effects of different treatment options (such as lithium salts and valproic acid). Indeed, studies investigating the issue in pediatric populations consistently show volumetric amygdalar decreases and so do research papers on first episode BD [67]. As we have proposed elsewhere, [78] molecular and cellular desensitization to glutamate as a tentative homeostatic adaptation to increased glutamatergic transmission could explain most of the findings: a chronic hyperactivity of amygdaloid structures, be it related to childhood stress or the allostatic load of mood episodes, could lead to neurotoxicity and glial cell loss. Conversely, mood stabilizers could exert neuroprotective effects on limbic structures, reducing overexcitation and normalizing receptor sensitivity. This interpretative strategy could partly resolve the outwardly contradictions found in the available scientific literature on this topic.

5. Conclusions

The relationship between the clinical expression of a disorder and its anatomo-functional correlates remains pivotal for a real understanding of the complex relations between brain and behavior [79, 80]. The disconnection paradigm, as envisaged by Geschwind in his landmark paper and revitalized today by the availability of methods for mapping connections and cortical activities in-vivo, seems to be of key importance to elaborate a comprehensive approach to the study of the brain-behavior relation. All the evidences reported in this chapter suggest that social behavior, learning, and memory depend on the activity of different brain cortical regions coordinated by subcortical structures, and that "emotions," whose effector nuclear complex is the AB, are the key starting points for all of these physiological processes. Concurrently, there have been numerous pieces of evidence that neuropsychiatric disorders related with mnesic functions, such as trauma spectrum and mood spectrum disorders, subtend an altered connectivity linked to an abnormal activity of the AB, [4, 12–17, 50, 51, 58–73, 77] and that its normalization by either pharmacological or non-pharmacological techniques could feasibly lead to a better clinical outcome [18–26, 45, 52, 54–57, 68, 74–76, 78]. In this scenario, the disconnection paradigm provides a renewed mindset to rethink neuropsychiatry and psychotherapy.

Author details

Alessandro Weiss[1]* and Francesco Weiss[2]

1 PerFormat Salute, Livorno, Italy

2 University of Pisa, Pisa, Italy

*Address all correspondence to: alessandroweiss1981@gmail.com

References

[1] Kandel ER, Schwartz JH, Jessel TM, Siegelbaum SA, Hudspet AJ. Principles of Neural Science. 5th ed. New York, USA: McGraw Hill; 2013. p. 1709

[2] Ellemberger HF. The Discovery of Unconscious. The History and Evolution of Dynamic Psychiatry. New York, USA: Basic Books; 1970

[3] Damasio A. Descarte's Error: Emotion, Reason and the Human Brain. New York, USA: Avon; 1994

[4] Nieuwenhuys R, Voogd J, can Huijzen C. The Human Central Nervous System. New York: Springer-Verlag; 2005

[5] Herrick CJ. The Evolution of Human Nature. Austin: University of Texas; 1956

[6] MacLean P. The Triune Brain in Evolution. Role in Paleocerebral Functions. New York, USA: Plenum; 1990

[7] Heimer L, Wilson RD. The subcortical projections of allocortex: Similarities in the neuronal associations of the hippocampus, the piriform cortex and the neocortex. In: Santini M, editor. Golgi Centennial Symposium Proceedings. New York: Raven Press; 1975. pp. 173-193

[8] Penfield W. Functional localization in temporal and deep sylvian areas. Research Publications-Association for Research in Nervous and Mental Disease. 1958;**36**:210-226

[9] Penfield W, Perot P. The brain's record of auditory and visual experience: A final summary and discussion. Brain. 1963;**86**:595-696

[10] Milner B, Squire LR, Kandel ER. Cognitive neuroscience and the study of memory. Neuron. 1998;**20**:445-468. DOI: 10.1016/s0896-6273(00)80987-3

[11] Schacter DL. Implicit memory: History and current status. Journal of Experimental Psychology: Learning, Memory, and Cognition. 1987;**13**:501-518

[12] Heimer L, Van Hoesen GW. The limbic lobe and its output channels: Implications for emotional functions and adaptive behaviour. Neuroscience and Biobehavioral Reviews. 2006;**30**(2):126-147. DOI: 10.1016/j.neubiorev.2005.06.006

[13] McGaugh JL. The amygdala modulates the consolidation of memories of emotionally arousing experiences. Annual Review of Neuroscience. 2004;**27**(1):1-28. DOI: 10.1146/annurev.neuro.27.070203.144157

[14] Gurvits TV, Shenton ME, Hokama H, Ohta H, Lasko NB, Gilbertson MW, et al. Magnetic resonance imaging study of hippocampal volume in chronic, combat-related posttraumatic stress disorder. Biological Psychiatry. 1996;**40**:1091-1099. DOI: 10.1016/S0006-3223(96)00229-6

[15] Schumann CM, Amaral DG. Sterreological analysis of amygdala neuron number in autism. The Journal of Neuroscience. 2006;**26**(29):7674-7679. DOI: 10.1523/JNEUROSCI.1285-06.2006

[16] Bouchard TP, Malykhin N, Martin WR, Hanstock CC, Emery DJ, Fisher NJ, et al. Age and dementia-associated atrophy predominates in the hippocampal head and amygdala in Parkinson's disease. Neurobiology of Aging. 2008;**29**:1027-1039. DOI: 10.1016/j.neurobiolaging.2007.02.002

[17] Baur V, Hänggi J, Langer N, Jäncke L. Resting-state functional and structural connectivity within an insula-amygdala route specifically index state and trait anxiety. Biological Psychiatry. 2013;**73**:85-92. DOI: 10.1016/j.biopsych.2012.06.003

[18] Kennedy SH, Giacobbe P, Rizvi SJ, Placenza FM, Nishikawa Y, Mayberg HS, et al. Deep brain stimulation for treatment-resistant depression: Follow-up after 3 to 6 years. The American Journal of Psychiatry. 2011;**168**:502-510. DOI: 10.1176/appi.ajp.2010.10081187

[19] Langevin JP. The amygdala as target for behavioural surgery. Surgical Neurology International. 2012;**3**(Suppl1):S40-S46. DOI: 10.4103/2152-7806.91609

[20] Schlaepfer TE, Bewernick BH, Kayser S, Mädler B, Coenen VA. Rapid effects of deep brain stimulation for treatment-resistant major depression. Biological Psychiatry. 2013;**73**(12):1204-1212. DOI: 10.1016/j.biopsych.2013.01.034

[21] Sturm V, Fricke O, Buhrle CP, Lenartz D, Maarouf M, Treuer H, et al. DBS in the basolateral amygdala improves symptoms of autism and related self-injurious behaviour: A case report and hypothesis on the pathogenesis of the disorder. Frontiers in Human Neuroscience. 2013;**6**:341. DOI: 10.3389/fnhnm.2012.00341

[22] Koek RJ, Langevin JP, Krahl SE, Kosoyan HJ, Schwartz HN, Chen JW, et al. Deep brain stimulation of the basolateral amygdala for treatment-refractory combat post-traumatic stress disorder (PTSD): Study protocol for a pilot randomized controlled trial with blinded, staggered onset of stimulation. Trials. 2014;**15**:356. DOI: 10.1186/1745-6215-15-356

[23] Langevin JP, Koek RJ, Schwartz HN, Chen JW, Sultzer DL, Mandelkern MA, et al. Deep brain stimulation of the basolateral amygdala for treatment-refractory posttraumatic stress disorder. Biological Psychiatry. 2015;**79**(10):e82-e84. DOI: 10-1016/j.biopsych.2015.09.003

[24] Luyten L, Hendrickx S, Raymaekers S, Gabriëls L, Nuttin B. Electrical stimulation in the bed nucleus of the stria terminalis alleviates severe obsessive-compulsive disorder. Molecular Psychiatry. 2016;**21**(9):1272. DOI: 10.1038/mp.2015.124

[25] Blomstedt P, Naesström M, Bodlund O. Deep brain stimulation in the bed nucleus of the stria terminalis and medial forebrain bundle in a patient with major depressive disorder and anorexia nervosa. Clinical Case Reports. 2017;**5**(5):679-684. DOI: 10.1002/ccr3.856

[26] Lavano A, Guzzi G, Della Torre A, Lavano SM, Tiriolo R, Volpentesta G. DBS in treatment of post-traumatic stress disorder. Brain Sciences. 2018;**8**(1) Pii:E18. DOI: 10.3390/brainsci8010018

[27] Weiss A, Di Carlo DT, Di Russo P, Weiss F, Castagna M, Cosottini M, et al. Microsurgical anatomy of the amygdaloid body and its connections. Brain Structure and Functions. 2021;**226**(3):861-874. DOI: 10.1007/s00429-020-02214-3

[28] Burdach KF. Vom Baue undLeben des Gehirns. Vol. 3. Leipzig: Dyk'sche Buchhdl; 1819-1826

[29] Meyer A. Karl Friedrich Burdach and his place in the history of neuroanatomy. Journal of Neurology, Neurosurgery, and Psychiatry. 1970;**33**(5):553-561. DOI: 10.1136/jnnp.33.5.553

[30] Klingler J, Gloor P. The connections of the amygdala and of the anterior

temporal cortex in the human brain. The Journal of Comparative Neurology. 1960;**115**:333-369. DOI: 10.1002/cne.901150305

[31] Di Marino V, Etienne Y, Niddam M. The Amygdaloid Nuclear Complex. Anatomic Study of the Human Amygdala. Marseille: Springer International Publishing AG Switzerland; 2016. p. 147. DOI: 10.1007/978-3-319-23243-0

[32] Dejerine J. Anatomie des centres nerveux. Vol. I. Paris: J. Rueff et Cie; 1895

[33] Turner BH, Knapp ME. Projections of the nucleus and tracts of the stria terminalis following lesions at the level of the anterior commissure. Experimental Neurology. 1976;**51**(2):468-479. DOI: 10.1016/0014-4886(76)90270-3

[34] Miller EJ, Saint Marie LR, Breier MR, Swerdlow NR. Pathways from the ventral hippocampus and caudal amygdala to forebrain regions that regulate sensorimotor gating in the rats. Neuroscience. 2010;**165**(2):601-611. DOI: 10.1016/j.neuroscience.2009.10.036

[35] Choi CY, Han SR, Yee GT, Lee CH. Central core of the cerebrum. Journal of Neurosurgery. 2011;**114**(2):463-469. DOI: 10.3171/2010.9.JNS10530

[36] Alarcon C, de Notaris M, Palma K, Soria G, Weiss A, Kassam A, et al. Anatomic study of the central core of the cerebrum correlating 7-T magnetic resonance imaging and fiber dissection with the aid of a neuronavigation system. Neurosurgery. 2014;**10**(2):294-304. DOI: 10.1227/NEU.0000000000000271

[37] Güngör A, Baydin S, Middlebrooks EH, Tanriover N, Isler C, Rhoton ALJR. The white matter tracts of the cerebrum in ventricular surgery and hydrocephalus. Journal of Neurosurgery. 2017;**126**(3):945-971. DOI: 10.3171/2016.1.JNS152082

[38] Serra C, Akeret K, Maldaner N, Staartjes VE, Regli L, Balsavias G, et al. A white matter fiber microdissection study of anterior perforated substance and the basal forebrain: A gateway to the basal ganglia? Operative Neurosurgery. 2019;**17**(3):311-320. DOI: 10.1093/ons/opy345

[39] Coenen VA, Panksepp J, Hurwitz TA, Urbach H, Madler B. Human medial forebrain bundle (MFB) and anterior thalamic radiation (ATR): Imaging of two major subcortical pathways and the dynamic balance of opposite affects in understanding depression. The Journal of Neuropsychiatry and Clinical Neurosciences. 2012;**24**(2):223-236. DOI: 10.1176/appi.neuropsych.11080180

[40] Rosene DL, Van Hoesen GW. Hippocampal efferents reach widespread areas of the cerebral cortex and amygdala in the rhesus monkey. Science. 1977;**198**(4314):315-317. DOI: 10.1126/science.410102

[41] Kemppainen S, Jolkkonen E, Pitkänen A. Projections from the posterior cortical nucleus of the amygdala to the hippocampal formation an parahippocampal region in rat. Hippocampus. 2002;**12**(6):735-755. DOI: 10.1002/hipo.10020

[42] Kemppainen S, Pitkänen A. Damage to the amygdalo-hippocampal projection in temporal lobe epilepsy: A tract-tracing study in chronic epileptic rats. Neuroscience. 2004;**126**(2):485-501. DOI: 10.1016/j.neuroscience.2004.03.015

[43] Curran EJ. A new association fiber tract in the cerebrum (with remarks on the fiber tract dissection method of studying the brain). Neuroscience. 1909;**19**:645657

[44] Monakow C. Experimentelle und pathologisch-anatomische Untersuchungen iiber die Haubenregion, den Sehiigel und die Regio Subthalamica, nebst Beitragen zur Kenntnis friih erworbener Gross-und Kleinhirndefekte. Archiv für Psychiatrie und Nervenkrankheiten. 1895;**27**:1-128; 386-478

[45] Jimenez F, Velasco F, Salìn-Pascual R, Velasco M, Nicolini H, Velasco AL, et al. Neuromodulation of the inferior thalamic peduncle for major depression and obsessive compulsive disorder. Acta Neurochirurgica Supplement. 2007;**97**(Pt2):393-398. DOI: 10.1007/978-3-211-33081-4_44

[46] Proust M. In Search of Lost Time. Vol. 6. London, UK: Allen Lane; 2002

[47] Bajada CJ, Banks B, Lambon Ralph MA, Cloutman LL. Reconnecting with Joseph and Augusta Dejerene: 100 years on. Brain. 2017;**140**(10):2752-2759. DOI: 10.1093/brain/awx225

[48] Heilbronner SR, Haber SN. Frontal cortical and subcortical projections provide a basis for segmenting the cingulum bundle: Implications for neuroimaging and psychiatric disorders. The Journal of Neuroscience. 2014;**34**(30):10041-11154. DOI: 10.1523/JNEUROSCI.5459-13.2014

[49] Peltier J, Verclytte S, Demaire C, Pruvo JP, Havet E, Le Gars D. Microsurgical anatomy of the anterior commissure: Correlations with diffusion tensor imaging fiber tracking and clinical relevance. Neurosurgery. 2011;**62**(2suppl):ons241-ons246. DOI: 10.1227/NEU.0b013e31821bc882

[50] de Olmos J, Heimer L. The concepts of the ventral striatopallidal system and extended amygdala. Annals of the New York Academy of Sciences. 1999;**29**(877):1-32. DOI: 10.1111/j.1749-6632.1999.tb09258.x

[51] Mayberg HS, Lozano AM, Voon V, McNeely HE, Seminowicz D, Hamani C, et al. Deep brain stimulation for treatment-resistant depression. Neuron. 2005;**45**:651-660. DOI: 10.1016/j.neuron.2005.02.014

[52] Mokhtari Hashtjini M, Pirzad Jahromi G, Meftahi GH, Esmaeili D, Javidnazar D. Aqueous extract of saffron administration along with amygdala deep brain stimulation promoted alleviation of symptoms in post-traumatic stress disorder (PTSD) in rats. Avicenna Journal of Phytomedicine. 2018;**8**(4):358-369

[53] Piacentiani S, Romito L, Franzini A, Granato A, Broggi G, Albanese A. Mood disorder following DBS of the left amygdaloid region in a dystonia patient with a dislodged electrode. Movement Disorders. 2008;**23**(1):147-150. DOI: 10.1002/mds.21805

[54] Steele JD, Christmas D, Eljamel MS, Matthews K. Anterior cingulotomy for major depression: Clinical outcome and relationship to lesion characteristics. Biological Psychiatry. 2008;**63**:670-677. DOI: 10.1016/j.biopsych.2007.07.019

[55] Tykocki T, Mandat T, Kornakiewicz A, Koziara H, Nauman P. Deep brain stimulation for refractory epilepsy. Archives of Medical Science. 2012;**8**(5):805-816. DOI: 10.5114/aoms.2012.31135

[56] Baydin S, Gungor A, Tanriover N, Rhothon AL Jr. Microsurgical and fiber tract anatomy of the nucleus accumbens. Operative Neurosurgery. 2016;**12**(4):E396-E397. DOI: 10.1227/NEU.0000000000001422

[57] Lee DJ, Dallapiazza RF, De Vloo P, Elias GJB, Fomenko A, Boutet A, et al.

Inferior thalamic peduncle deep brain stimulation for treatment-refractory obsessive-compulsive disorder: A phase 1 pilot trial. Brain Stimulation. 2019;**12**(2):344-352. DOI: 10.1016/j.brs.2018.11.012

[58] American Psychiatric Association, DSM-5 Task Force. Diagnostic and Statistical Manual of Mental Disorders. 5th ed. Washinton DC, USA: American Psychiatric Assocation; 2013. DOI: 10.1176/appi.books

[59] Van Der Kolk BA, Pelcovitz D, Roth S, Mandel R, McFarlane A, Herman J. Dissociation, somatization and affect dysregulation. The complexity of adaptation to trauma. American Journal of Psychiatry. 1996;**153**(7):89-93. DOI: 10.1176/ajp.153.7.83

[60] Van der Hart O, Nijenhuis ERS, Steele K. The haunted self. In: Structural Dissociation and The Treatment of Chronic Traumatization. New York, USA: Norton Professional Books; 2006

[61] Roeckner AR, Oliver KI, Lebois LAM, van Rooij Sanne JH, Stevens JS. Neural contributors to trauma resilience: A review of longitudinal neuroimaging studies. Translational Psychiatry. 2021;**11**(1):508. DOI: 10.1038/s41398-021-01633-y

[62] Sherin JE, Nemeroff CB. Post-traumatic stress disorder: The neurobiological impact of psychological trauma. Dialogues Clininical Neuroscience. 2011;**13**(3):263-278. DOI: 10.31887/DCNS.2011.13.2/jsherin

[63] Pagani M, Hogberg G, Fernandez I, Siracusano A. Correlates of EMDR therapy in functional and structural neuroimaging: A critical summary of recent findings. Journal of EMDR Practice and Research. 2013;7(1):29-38. DOI: 10.3389/fpsyg.2018.01395

[64] Porges SW. The polyvagal perspective. Biological Psychology. 2007;**74**(2):116-143. DOI: 10.1016/j.biopsycho.2006.06.009

[65] Boeckle M, Schrimpf M, Liegl G, Pieh C. Neural correlates of somatoform disorders from a meta-analytic perspective on neuroimaging studies. NeuroImage: Clinical. 2016;**10**(11):606-613. DOI: 10.1016/j.nicl.2016.04.001

[66] Oakes P, Loukas M, Oskouian RJ, Tubbs RS. The neuroanatomy of depression: A review. Clinical Anatomy. 2017;**30**(1):44-49. DOI: 10.1002/ca.22781

[67] Garrett A, Chang K. The role of the amygdala in bipolar disorder development. Development and Psychopathology. 2008;**20**(4):1285-1296. DOI: 10.1017/S0954579408000618

[68] Han KM, De Berardis D, Fornaro M, Kim YK. Differentiating between bipolar and unipolar depression in functional and structural MRI studies. Progress in Neuro-Psychopharmacology & Biological Psychiatry. 2019;**91**:20-27. DOI: 10.1016/j.pnpbp.2018.03.022

[69] Long X, Li L, Wang X, et al. Gray matter alterations in adolescent major depressive disorder and adolescent bipolar disorder. Journal of Affective Disorders. 2023;**325**:550-563. DOI: 10.1016/j.jad.2023.01.049

[70] Mourão-Miranda J, Almeida JRC, Hassel S, et al. Pattern recognition analyses of brain activation elicited by happy and neutral faces in unipolar and bipolar depression. Bipolar Disorders. 2012;**14**(4):451-460. DOI: 10.1111/j.1399-5618.2012.01019.x

[71] Grogans SE, Fox AS, Shackman AJ. The amygdala and depression: A sober reconsideration. The American Journal

of Psychiatry. 2022;**179**(7):454-457. DOI: 10.1176/appi.ajp.20220412

[72] Cotovio G, Oliveira-Maia AJ. Functional neuroanatomy of mania. Translational Psychiatry. 2022;**12**(1):29. DOI: 10.1038/s41398-022-01786-4

[73] Post RM. Transduction of psychosocial stress into the neurobiology of recurrent affective disorder. The American Journal of Psychiatry. 1992;**149**(8):999-1010. DOI: 10.1176/ajp.149.8.999

[74] Post RM, Weiss SR. A speculative model of affective illness cyclicity based on patterns of drug tolerance observed in amygdala-kindled seizures. Molecular Neurobiology. 1996;**13**(1):33-60. DOI: 10.1007/BF02740751

[75] Post RM. The status of the sensitization/kindling hypothesis of bipolar disorder. Current Psychosis & Therapeutics Reports. 2004;**2**(4):135-141. DOI: 10.1007/BF02629414

[76] Post RM, Denicoff KD, Frye MA, et al. A history of the use of anticonvulsants as mood stabilizers in the last two decades of the 20th century. Neuropsychobiology. 1998;**38**(3):152-166. DOI: 10.1159/000026532

[77] Strakowski SM, Adler CM, Almeida J, et al. The functional neuroanatomy of bipolar disorder: A consensus model. Bipolar Disorders. 2012;**14**(4):313-325. DOI: 10.1111/j.1399-5618.2012.01022.x

[78] Weiss F, Caruso V, De Rosa U, Beatino MF, Barbuti M, Nicoletti F, et al. The role of NMDA receptors in bipolar disorder: A systematic review. Bipolar Disord. 19 May 2023. DOI: 10.1111/bdi.13335. Epub ahead of print. PMID: 37208966

[79] Thiebaut de Schotten M, Dell'Acqua F, Ratiu P, Leslie A, Howells H, Cabanis E, et al. From Phineas Gage and Monsier Leborgne to H.M.: Revisiting disconnection syndromes. Cerebral Cortex. 2015;**25**(12):4812-4827. DOI: 10.1093/cercor/bhv173

[80] Mah YH, Husain M, Rees G, Nachev P. Human brain lesion-deficit inference remapped. Brain. 2014;**2014**(137):2522-2531. DOI: 10.1093/brain/awu164

Chapter 6

Olfactory Imagery and Emotional Control

Martin R. Portner

Abstract

Olfaction and gustation are important survival mechanisms. These sensory modalities also have an impact on memory and emotions. Olfactory stimulation has recently been used in virtual reality environments to treat emotional distress. There is evidence that olfactory and gustatory stimuli reach the insula, where they influence a number of other brain networks. There is little research on the use of smell and taste during mindfulness, but it will be shown that they can foster balanced emotional responses. In this chapter, we will look at how to incorporate olfactory and gustatory memory-based experiences during mindfulness exercises to bring about emotional homeostasis.

Keywords: smell, olfaction, taste, gustatory, mindfulness, olfactory virtual reality, insula

1. Introduction

It is true that the senses of smell and taste are often overlooked nowadays, but the fact remains that the mind relies heavily on olfaction and gustation to communicate with the body. Furthermore, certain scents and tastes can have a profound effect on moods and emotions, and these ancient sensory modalities have ensured humanity's survival. In addition to supporting our daily lives, they are used for the appreciation of perfume and wine, and scent-based therapies have been shown to be effective treatment for conditions such as anxiety [1] and depression [2].

The most powerful human sense may also be the least used. The scent is a sense rapidly activated when we are born and is also a fierce memory encoder during development. Science has repeatedly demonstrated how deeply the olfactory system is linked to an individual's emotional state [3]. Scent has been shown to have a connection with the brain areas responsible for motivation and decision-making [4], and it may emerge as a powerful strategic tool for changing our behavior and moods.

2. A painful experience

An interesting experiment has shown how common people unconsciously smell their hands [5]. Ninety-four percent of 400 respondents acknowledged engaging in smelling their hands. The study also showed that simultaneous motor expression was crucial for that because the nasal airflow more than doubled when the hands were at the nose, indicating that people were in fact sniffing their hands' smell.

Movement appears to be important during odor appreciation as reducing the sniffing movement of the nose impacts the degree of pleasantness of the smell [6]. Sniffing seems to happen in a certain order during olfaction, and it may serve the same purpose as moving the eyes during visual imagery.

Regarding why such an unconscious smell of the hands takes place, it appears that, similarly to what happens when one looks at herself/himself in the mirror, smelling one's hands could be related to obtaining a notion of one's self through olfaction. Thus, by sniffing their hands, humans may subconsciously obtain reassurance about themselves.

When most people think about the sense of smell or taste, they might imagine something like enjoying a fragrant flower or savoring a delicious meal. However, smell and taste may play key roles in emotional well-being. In this respect, it could be argued that smelling one's hands could in fact lead to emotional homeostasis—trying to return to the present moment when emotional balance swings.

It appears that a connection has been made between olfaction and pain. It has been shown that exposure to odors results in an elevated pain threshold in people who suffer from persistent low back pain [7]. How could such a process be explained? An experiment has shown that at the level of the mouse input trigeminal system, olfactory stimulation seems to modify nociceptive processing [8]. In the presence of pain, the floral odorants phenylethyl alcohol and lavender oil decreased the amount of network activity in the caudalis trigeminal nucleus. It was hypothesized that olfactory inputs excite hypothalamic networks, which in turn alter network activity in the trigeminal nucleus. It would appear that exposure to odors stimulates activity in the hypothalamus and inhibits the processing of painful stimuli, a form of descending inhibition of the pain modulation mechanism. The fact that smelling includes a significant hedonic component lends weight to the idea that it can be used as a technique for relieving pain. The authors suggest that such an approach could transform olfaction into a therapeutic tool for the management of headaches.

In his book "The Brain's Way of Healing: Remarkable Discoveries and Recoveries from the Frontiers of Neuroplasticity," Canadian psychiatrist and researcher Norman Doidge described the case of a woman who sustained a lower back injury with multiple lumbar vertebrae fractures [9]. After the traumatic event, she suffered from intractable pain for years, necessitating ever-increasing opioid medications. A new possibility emerged when she met Dr. Michael Moskowitz, a chronic pain specialist. Moskowitz had himself suffered a cervical injury and, after many years of increasing pain, resorted to attempting to ease his pain by trying to mentally modify the cerebral maps that underlined it. He thought that rewiring the brain's pain connections was possible, so he began to support brain neuroplasticity as a way to treat pain.

He explained to his patient that he was convinced, he had no other effective means to bring her pain relief, and he advised her to try the visualization technique he had used on himself. It was not an easy task, but with a persistent and focused approach, the patient began to realize she had some moments of feeling no pain. After some months, she was able to live for days without pain. Many months later, she was pain-free and, after years of monitoring, she was declared cured.

Expanding on the brain maps for pain, Doidge argues that the abnormal connections of the central nervous system circuits underlying chronic pain are brought together under the principle that neurons that fire together, wire together. Hence, if properly directed by visualization, such connections could be unlearned by progressively unwiring the pain connections.

Moskowitz discovered that in the use-it-or-lose brain, there is constant competition for cortical estate because the brain's activities take up more and more circuits by capturing resources from other areas. He drew three brain images: one in acute pain, with activity in 16 brain areas; one in chronic pain, with the same areas expanded; and one in which the brain registers no pain at all [9]. As a result, whenever Moskowitz felt pain, he immediately began visualizing his brain maps. To begin, he would visualize the image of the brain in chronic pain and note how much that map had expanded due to newly added neuroplastic neurons. Then, he imagined the firing areas shrinking until they resembled a painless image of the brain. He was aware, for instance, that the posterior cingulate cortex and posterior parietal lobes co-processed vision and pain. He knew that by visualizing these brain maps repeatedly, he was forcing their neurons to limit pain processing in order to accommodate visual inputs. He imagined; he was transferring pain to vision. His experience demonstrates that he achieved success. He had been in excruciating pain for 13 years before using his visualization technique and was pain-free in 6 weeks.

In point of fact, Moskowitz's strategy can be construed as an instance of practicing a specific form of mindfulness. The results of a study conducted on female college students who were experiencing stress, anxiety, and sadness and who participated in an 8-week program that combined visualization and meditation found that the combination of these two techniques increased self-awareness and decreased anxiety [10]. Pain relief was associated with orbitofrontal cortex and rostral anterior cingulate cortex regulation of low-level nociceptive neural targets (the thalamus and primary somatosensory cortex). Following mindfulness-based mental training, it appears that a reappraisal mechanism was activated, possibly by rewiring these neuronal circuits [11].

There is some evidence to suggest that olfactory inputs to the brain can alleviate physical pain by having an effect on the connections that house emotionally painful experiences. The amygdala, orbitofrontal cortex, and hippocampus, for example, are all involved in the process of odor-elicited emotion as well as in the recalling of odor-associated emotional memory [12]. The citrus fragrance has been shown to have a beneficial effect on the mental states and immune function of patients who are depressed [13]. It has also been demonstrated that rose essential oils prevent the disruption of the skin barrier and the elevation of salivary cortisol in humans caused by stress [14], and that the aromas of orange and lavender counteract the anxiety experienced by patients in a setting similar to that of a dental office [15].

The influence of gustation on both physical pain and emotional experience is much less well understood. When evaluated using heat detection thresholds and the perception of painfully hot stimuli, solutions of sucrose and sucralose exhibited no evidence of analgesic activity [16]. Positive emotions were associated with increased sweet and decreased sour intensities, while negative emotions were associated with increased sour and decreased sweet tastes. On the other hand, positive emotions were associated with increased sweet and decreased sour intensities [17].

3. Memory storage

Invoking positive autobiographical memories has been shown to be an effective therapeutic technique for reducing feelings of emotional distress in a wide variety of clinical conditions. In general, one could say that positive autobiographical memories have the ability to improve mood (for a review, see [18]). Odors may be particularly helpful for enhancing mood states because they elicit more profound emotional memories than other types of stimuli and because the memories that are evoked by odors are typically pleasant. It was discovered that men and women who were exposed to the scent of a perfume that personally reminded them of themselves experienced more positive emotions, an increased mood of happiness, and decreased anxiety when compared to when they were exposed to a pleasant fragrance that did not elicit personal memories.

In accordance with the principles of odor-associative learning, when an odor becomes linked to the emotional properties of an event, it can produce a downstream effect on physiology. A calming odor, for example, will also slow heart rate and respiration rate, as well as improve immune responses [19]. On the other hand, an odor that evokes an important event from one's past emotional history can cause one's heart to race and a jolt of adrenaline to be released. Because of the direct link between olfactory processing and the amygdala-hippocampal complex, emotions elicited by odor-evoked memories can arise immediately upon perception. An odor-evoked memory can thus be defined as possessing emotional primacy, which occurs when affect is elicited before a cognitive understanding of why that emotion occurred in the first place.

Odor-evoked memories could thus elicit specific emotions such as self-confidence, motivation, and energizing behavior. All of these states, for example, could be triggered by an odor that evokes the memory of winning a significant event, resulting in positive physiological effects [20]. Considering that olfactory cognition is not affected by retroactive interference, a specific odor associated with a significant past personal event can be considered a potential therapeutic agent.

Because odor-evoked memories are emotionally charged, they recall our past in a visceral way more than any other type of memory experience. This exceptional feature of olfaction can be seen in people who suddenly lost their sense of smell, taste, or both during the COVID-19 pandemic. Olfactory training has been shown to be the only disease-specific intervention with evidence of efficacy for the treatment of postinfectious olfactory dysfunction. Treatment based on sniffing and trying to identify scents, such as rose, eucalyptus, lemon, and clove, over the course of several months should be implemented as soon as possible [21].

It has been reported that emotional distress caused by the COVID-19 pandemic is expected to be a major mental health issue in the coming years, including post-traumatic stress disorder (PTSD). Perhaps a remarkable example of using odor treatment for emotional distress is its use to treat patients with PTSD. In recent years, we have seen the emergence of olfactory virtual reality (OVR) as an effective means to treat PTSD [22, 23].

Virtual reality (VR) technology is becoming more prevalent in healthcare settings. It provides an experience that is unrivaled in real life. VR applications have the potential to transform the assessment, understanding, and treatment of mental health problems, and have been used in studies to treat pain in patients undergoing treatments such as chemotherapy [24], as well as emotional disorders, such as stress and low self-efficacy [25].

When VR alters other senses, such as sight and spatial perception, olfactory stimuli can reconnect oneself with the other senses. Olfactory stimuli can influence psychosocial functioning by altering negative emotional states such as hostility, jitteriness, upset, and distress. Reduced negative effect improves these emotional responses. Because confidence is increased, improvements in self-efficacy are frequently observed.

Odors are common and debilitating triggers of distressing emotions, and OVR technology is being adapted to treat such disorders. It appears to be uniquely positioned to be used in the treatment of PTSD symptoms due to unique emotional features of olfactory processing and odor-evoked memory. Systematic desensitization, prolonged exposure, reappraisal strategies, and conditioning odors are being used for emotional calming and are examples of OVR treatment strategies for PTSD [22].

Preventing PTSD in vulnerable groups can be accomplished by neutralizing potential odor triggers, such as burned fuel or flesh, through preemptive habituation and reappraisal training. In addition to PTSD, OVR has shown promise in treating other types of emotional problems as well. In a test for improving psychosocial well-being with OVR, for example, the presence of a pleasant odor significantly reduced negative affect on several measures, such as hostility, anxiety, and distress, compared to when there was no odor. OVR is a computer-simulated, three-dimensional environment where the user wears a headset that blocks out most of the outside world and replaces it with new information. It is the experience of presence—the illusion of "being there"—in a world that exists outside of the self [26], where the external physical environment disappears from the user's phenomenal awareness. It is created by a headset that provides its own sensory and psychological landscape. The desired state in OVR is presence, a complex psychological experience formed by the multifaceted interaction between sensory stimulation, including smell, and the resulting cognitive responses. Also, the person can interact with the virtual environment and become "immersed" in it by using hand controls or other interface devices [27].

PTSD is a severely debilitating emotional disorder characterized by extremely distressing thoughts and intrusive traumatic memories. Depression, substance abuse, sleep problems, and anxiety are all common symptoms. According to the American Psychiatric Association, it affects approximately 3.5% of adults in the United States, with one in every 11 people suffering from it at some point in their lives [28]. Notably, alarmingly high rates of PTSD have been reported since the COVID-19 pandemic began, and there is a looming mental health crisis among healthcare workers and the general public. As a result, PTSD and all other forms of anxiety, perhaps even depression, will likely affect a much larger proportion of the population than they did prior to 2020, and rates are expected to rise over the next few years.

4. Odor and tastes return

Odors with positive emotional associations can be used to repair unpleasant emotional states and prevent unhealthy behaviors [29]. Subjectively pleasant odors, for example, can improve mood, increase nostalgia and its psychological benefits [30], reduce cigarette cravings in smokers [31], and decrease cravings for unhealthy foods [32]. An odor that induces calming emotions will also slow heart rate and respiration rate, as well as improve immune responses. Based on these principles, positive emotional associations with odors can be used to reduce the unpleasantness of traumatic memories. If a person is emotionally overwhelmed by a traumatic recollection, using a technique based on odor mindfulness can induce relaxation, de-escalate emotions, and change the mindset.

As exposure to an olfactory stimulus has been shown to reduce pain, improve negative affect and increase emotional resilience, it opens up an avenue to question whether these effects can be obtained through an odor *willfully recalled* from memory.

For many years, it was assumed that odor could not be perceived when the primary chemosensory substance was absent. However, like visual mental imagery, olfactory mental imagery shares many neural processes with true olfactory perception. Methods from neuropsychology [33] and neuroimaging [34] have revealed striking similarities between visual perception and visual imagery. When positron emission tomography is used, it reveals that odor mental imagery is associated with the activation of odor-processing regions such as the piriform and orbitofrontal cortices [35].

Not only does the activity elicited by imagining odors mirror that is elicited by perceiving real odorants in specific brain regions but it also has a hedonic-specific pattern [36]. Its net effect is that during mental imagery of olfactory events, primary and secondary sensory cortical structures are activated and then expanded to emotion-processing brain centers.

It is important to emphasize that olfaction integrates a sensory and a motor component (the sniff for olfaction). It appears that olfactory perception is not simply passively induced but includes the reenacting of some form of motor input. Therefore, sniffing, the motor component of olfaction, is likely to be a prominent feature of olfactory reenactment [37].

Olfactory mental images can be generated from long-term memories. However, information in long-term memory is implicit and poorly accessible, as opposed to information in short-term memory, which is explicit and accessible. Therefore, sniffing during olfactory imagery is likely to activate an internal representation stored in long-term memory; this is used to generate the mental image, which includes a sensory-type representation in short-term memory. Sniffing while recalling an odor makes the information explicit and thus accessible.

If remembering an odor from memory activates the very same cortical areas that smelling itself does, and considering olfaction influences the emotional behavior of humans, could we use mental imagery of odors as a potential means to target feelings and behavior? In order to provide a response to this question, I will first examine how odor imagery can influence activity in the olfactory and nonolfactory regions of the brain.

5. The power of the insula

The primary olfactory cortex (piriform cortex), the secondary olfactory cortex (posterior orbitofrontal cortex), and the insula have all been shown to be activated during odor imagery. Perfumers have larger gray-matter volumes in areas associated with olfactory processing, such as the insula and anterior cingulate, which is an important region for emotional and cognitive integration [38].

A thicker and denser insular cortex has been shown to be associated with better olfactory performance in healthy people [39], while patients with various degrees and forms of olfactory dysfunction show loss of insular gray matter [40]. It is also well established that both olfactory and gustatory sensory modalities project to the prefrontal and insular cortices, and both odor and gustatory stimuli can exert calming effects when an olfactory or gustatory stimulus is presented [41, 42].

Mindfulness meditation has been shown to be capable of treating pain and emotional-related disorders [43–45], though its effect in the long run still remains to be elucidated. Mindfulness is a self-administered reflexive technique that has an impact on several brain structures. It is capable of increasing blood flow to the insula [46] and enriching the gray-matter volume in the insular cortex [47].

More recently, a groundbreaking approach to treating chronic pain has been advanced, whereby patients seek to *willfully activate* a particular brain region while receiving real-time functional magnetic resonance imaging. It was shown that participants could voluntarily self-activate the cingulate and insula using real-time functional magnetic resonance imaging (rtf-MRI) and that such activation resulted in pain reduction [48] as well as a better balance between stress-involved brain structures [49]. It was clear that when given feedback about their performance with rtf-MRI, participants were able to learn a strategy to control their minds.

Participants in such studies were able to change neuronal processing within the insula, but the precise mechanism by which this was accomplished is still unknown. It has been demonstrated in normal subjects [49] as well as in a small group of patients with a psychopathic disorder [50]. The anterior insula was clearly regulated by one criminal psychopathic individual. This participant not only learned to up-regulate the left anterior insular cortex, increasing the number of connections in the emotional network, but he also learned to increase the insula's difference between outgoing and incoming connections.

The brain insula is a region that possesses connectivity to various other brain centers. It is involved in interception and decision-making, connecting worldly events to inner states, and everyday activities such as pain, love, emotion, craving, addiction, music enjoyment, or even wine tasting. The insula may be partly responsible for all of these seemingly disparate things because it lies at the center of the concept of self-awareness—the fact that humans can be aware of themselves, their bodies, and their emotions, and motivate actions that define the present moment [51].

According to Damasio's [52] somatic marker hypothesis, people use bodily signals to help them make decisions. The insula uses its connections to influence decision-making because it is involved in the processing of these bodily sensations. Because of its constant bombardment with information about the body's location, condition, subjective emotions, and key features of its environment, the insula plays an important role in the foundation of our overall awareness. This compiled data is incorporated into what Craig refers to as a "global emotional moment" (51), a snapshot of ourselves at a single point in time that contains all of the information that is important to us—for example, being happy while hungry. It is the insula's role in stringing together these global emotional moments that allow us to be aware of the day's ongoing moments.

It has been shown that the insula can be a target region for rtf-MRI-based treatment for patients with emotional disorders. Further, this part of the brain can be activated consciously with feedback or without it. It has been shown to be activated during real time functional magnetic resonance imaging [53] and also when biofeedback was subsequently removed [54]. So, it would be reasonable to think that reenacting smelling and tasting may activate the insula in a way not based on feedback.

Odor and gustatory recollection from memory both activate the insula. Hence, I propose the mechanism of a switch, whereby recalling an odor or having a gustatory-based sensory experience activates the insula and brings us back to the present moment. During a mindfulness exercise, reenacting stored interoceptive (smell and gustation) and exteroceptive (sniffing and tasting) information into working memory engages the insula. For instance, mindfully reenacting a smell or taste or both could be brought about at a time of the day when one falls prey to negative thoughts or bad feelings. As the insula is shifted to the present moment, rumination is stopped by the recall of the pleasant olfactory and gustatory imagery. Furthermore, if practiced repeatedly, it could stimulate the insular cortical system, leading to neuroplasticity. Learning to mindfully recollect olfaction and gustation could be envisioned as a way to facilitate living in the present moment, increasing emotional appraisal, and facilitating decision-making.

6. Focused mindfulness

Mindfulness could be viewed as a method for activating the insula, whether through real time function neuroimaging, after-feedback learned to control, or mindful olfactory-gustatory recall. Borrowing from Kabat-Zinn, mindfulness could be defined as awareness that arises from activating the insula by paying attention on purpose, in the present moment, and non-judgmentally [55]. The mindful reenactment of a memory odor or flavor redirects insular resources away from pain, emotional rumination, and helplessness.

It is also possible that repeated application of this gustatory or smell-based mindfulness technique will result in increased insular neuroplasticity. It has been shown, for example, that olfactory perception regulates the olfactory brain's experience-dependent neuroplastic properties. This mechanism underpins the acquisition of fine-grained percepts that distinguish the perfume of Rosa damascena (Bulgarian Rose) from that of Rosa centifolia (Rose Maroc), allowing us to appreciate the wide range of aromas encountered in daily life [56].

It has also been demonstrated that olfactory training induces changes in functional connectivity in the left and right piriform cortices in anosmic patients with long-term smell loss due to infection before and after a 12-week period of olfactory training [57]. Prior to olfactory training, a broad network encompassing predominantly nonolfactory regions, such as the prefrontal areas, left inferior frontal gyrus, and left premotor cortex, was active; following training, these nonolfactory functional linkages were reduced. This suggests that after olfactory training, a process of brain remodeling occurs. It may play a similar role in mindfulness-based exercises involving gustatory or olfactory reenactments.

Only recently has science begun to comprehend how intimately the olfactory and gustatory systems are tied to an individual's emotional state. It has been demonstrated that smell and taste influence regions of the brain involved in emotion and motivation, and that actions, such as what we eat and drink, affect our daily productivity. Consequently, it could be a formidable strategic instrument for modifying behavior and emotions.

Does this suggest that we can achieve a runner's high or a state of mindfulness merely by recalling odors or tastes? Perhaps. The limbic system, which is the portion of the brain responsible for smell and emotion, contributes to a wide range of processes, including motivation, long-term memory, behavior, and, of course, smell. However, the most important function of the limbic system is to set the tone for our emotional self, altering throughout the day to generate the appropriate energy and emotions to execute tasks and preserve memories.

During this time, the olfactory and gustatory regions engage directly with the hippocampus in order to recover memories assisted by smell or taste. Consider the rush of recollections that accompanies getting a whiff of a loved one's scent or the aroma emanating from a glass of wine reminiscent of a memorable meeting.

Smelling and tasting are sensory experiences that imprint in memory the sense of well-being felt at the time they occur. We know that these memories can be recovered whenever a sensory stimulation occurs by chance; evidence demonstrates that we can act on them by deliberately recalling fragrances or tastes, giving us the opportunity to be touched by the initial emotions.

The insula, which is concealed within the lateral sulcus of the human brain, has a unique functional profile at the intersection of the internal and external environment, the crossroads of cortical and subcortical neural hierarchies, and as a functional lever at the intersection of the major large-scale functional brain networks. The insula gets a large number of cross-modal afferents and is best viewed as a multimodal integration site [58]. As it is connected to the limbic system, the reward system, and the frontal cortex, which are all engaged in cognitive, affective, and executive activities, it lays the groundwork for a rich functioning experience.

The hypothesis that recalling memories of gustatory and olfactory experiences during focused mindfulness strengthens the insula's role as a functional switch, leading to greater executive and emotional control, no doubt, needs to be tested further. Meanwhile, we could just be a step back from demonstrating that the pleasure of living one's life to the fullest is only a sniff or a taste away.

7. Conclusion

Smell and taste are complex mental experiences, and like hearing and vision, they can be reenacted. You can shut your eyes and visualize the music concert you attended. You can also vividly hear the chord progression of your favorite tune without any music in the background. Similarly, the smells of baked cake, coffee, or wine, as well as the tastes of barbecue, cheese, and beer, can be recreated. These odor and taste reenactments have the property of quickly building a bridge from a past experience to appreciation in the present moment, leaving ruminating unwanted thoughts or unpleasant feelings at bay. Sniffing and tasting have been important for evolution. Maybe we kept them in our memories as a way to help us get through hard times.

Conflict of interest

The author declares no conflict of interest.

Author details

Martin R. Portner
Instituto do Cérebro e Coluna, Gramado, RS, Brazil

*Address all correspondence to: mrportner@gmail.com

References

[1] Sánchez-Vidaña DA, Ngai SP, He W, Jk C, Lau BW, Tsang HW. The effectiveness of aromatherapy for depressive symptoms: A systematic review. Evidence-Based Complementary and Alternative Medicine. 2017;**2017**:586

[2] Barati F, Nasiri A, Akbari N, et al. The effect of aromatherapy on anxiety in patients. Nephro Urology Monthly. 2016;**8**:e38347

[3] Mouly AM, Sullivan R. Memory and plasticity in the olfactory system: From infancy to adulthood. In: Menini A, editor. The Neurobiology of Olfaction. Boca Raton: CRC Press/Taylor & Francis; 2010

[4] Shanahan LK, Bhutani S, Kahnt T. Olfactory perceptual decision-making is biased by motivational state. PLoS Biology. 2021;**19**:e3001374

[5] Perl O, Mishor E, Ravia A, Ravreby I, Sobel N. Are humans constantly but subconsciously smelling themselves? Philosophical Transaction on Royal Society. 2020;**375**:20190372

[6] Bensafi M, Porter J, Pouliot S, Mainland J, Johnson B, Zelano C, et al. Olfactomotor activity during imagery mimics that during perception. Nature Neuroscience. 2003;**6**:1142-1144

[7] Gossrau G, Baum D, Koch T, Sabatowski R, Hummel T, Haehner A. Exposure to Odors increases pain threshold in chronic low back pain patients. Pain Medicine. 2020;**21**:2546-2551

[8] Papotto N, Reithofer S, Baumert K, Carr R, Möhrlen F, Frings S. Olfactory stimulation inhibits nociceptive signal processing at the input stage of the central trigeminal system. Neuroscience. 2021;**479**:35-47

[9] Doidge N. The Brain's Way of Healing: Remarkable Discoveries and Recoveries From the Frontiers of Neuroplasticity. New York: Viking; 2015

[10] Margolin I, Pierce J, Wiley A. Wellness through a creative lens: Meditation and visualization. Journal of Religion & Spirituality in Social Work: Social Thought. 2011;**30**:234-252

[11] Zeidan F, Emerson NM, Farris SR, Ray JN, Jung Y, McHaffie JG, et al. Mindfulness meditation-based pain relief employs different neural mechanisms than placebo and sham mindfulness meditation-induced analgesia. The Journal of Neuroscience. 2015;**35**:15307-15325

[12] Kadohisa M. Effects of odor on emotion, with implications. Frontiers in Systems Neuroscience. 2013;**7**:1-6

[13] Komori T, Fujiwara R, Tanida M, Nomura J, Yokoyama MM. Effects of citrus fragrance on immune function and depressive states. Neuroimmunomodulation. 1995;**2**: 174-180

[14] Fukada M, Kano E, Miyoshi M, Komaki R, Watanabe T. Effect of "rose essential oil" inhalation on stress-induced skin-barrier disruption in rats and humans. Chemical Senses. 2012;**37**:347-356

[15] Lehrner J, Marwinski G, Lehr S, Johren P, Deecke L. Ambient odors of orange and lavender reduce anxiety and improve mood in a dental office. Physiology & Behavior. 2005;**86**:92-95

[16] Mooney ER, Davies AJ, Pickering AE. Sweet taste does not modulate pain perception in adult humans. Wellcome Open Research. 2020;**5**:43

[17] Noel C, Dando R. The effect of emotional state on taste perception. Appetite. 2015;**95**:89-95

[18] Herz RS. The role of odor-evoked memory in psychological and physiological health. Brain Sciences. 2016;**6**:22

[19] Matsunaga M, Isowa T, Yamakawa K, Kawanishi Y, Tsuboi H, Kaneko H, et al. Psychological and physiological responses to odor-evoked autobiographical memory. Neuro Endocrinology Letters. 2011;**32**(6):774-780

[20] Engen T. The acquisition of odor hedonics. In Perfumery: The Psychology and Biology of Fragrance. New York: Chapman & Hall; 1988

[21] Boscolo-Rizzo P. Smell and taste dysfunction after covid-19. BMJ. 2022;**378**

[22] Herz RS. Olfactory virtual reality: A new frontier in the treatment and prevention of posttraumatic stress disorder. Brain Sciences. 2021;**11**:1070

[23] Freeman D, Reeve S, Robinson A, Ehlers A, Clark D, Spanlang B, et al. Virtual reality in the assessment, understanding, and treatment of mental health disorders. Psychological Medicine. 2017;**47**:2393-2400

[24] Chirico A, D'Aiuto M, Pinto M, Milanese C, Napoli A, Avino F, et al. The elapsed time during a virtual reality treatment for stressful procedures. A pool analysis on breast cancer patients during chemotherapy. In: Pietro G, Gallo L, Howlett R, Jain L, editors. Smart Innovation, Systems and Technologies. Cham: Springer; 2016

[25] Kaimal G, Carroll-Haskins K, Ramakrishnan A, Magsamen S, Arslanbek A, Herres J. Outcomes of visual self-expression in virtual reality on psychosocial well-being with the inclusion of a fragrance stimulus: A Pilot Mixed-Methods Study. Frontiers in Psychology. 2020;**11**:589461

[26] Steuer J. Defining virtual reality: Dimensions determining telepresence. The Journal of Communication. 1992;**42**:73-93

[27] Spagnolli A, Gamberini L. A Place for presence: Understanding the human involvement in mediated interactive environments. Psychology Journal. 2005;**3**:6-15

[28] American Psychiatric Association. New APA Poll Shows Sustained Anxiety among Americans; More than Half of ParentsAre Concerned about the Mental Well-Being of Their Children. Available online: https://www.psychiatry.org/newsroom/news-releases/new-apa-poll-shows-sustained-anxiety-among-americans-more-than-half-of-parents-are-concerned-aboutthe-mental-well-being-of-their-children

[29] Herz RS. Mood and cognitive performance during odor exposure. In: Rouby C, Schaal B, Dubois D, Gervais R, Holley A, editors. Olfaction, Taste and Cognition. New York: Cambridge University Press; 2002. pp. 160-177

[30] Reid CA, Green JD, Wildschut T, Sedikides C. Scent-evoked nostalgia. Memory. 2015;**23**:157-166

[31] Sayette MA, Marchetti M, Herz RS, Martin LM, Bowdring MA. Pleasant olfactory cues can reduce

cigarette craving. Journal of Abnormal Psychology. 2019;**128**:327-340

[32] Kemps E, Tiggemann M. Olfactory stimulation curbs food cravings. Addictive Behaviors. 2013;**38**:1550-1554

[33] Farah MJ. Is visual imagery really visual? Overlooked evidence from neuropsychology. Psychological Review. 1988;**95**:307-317

[34] Kosslyn S. Understanding the mind's eye...and nose. Nature Neuroscience. 2003;**6**:1124-1125

[35] Djordjevic J, Zatorre RJ, Petrides M, Boyle JA, Jones-Gotman M. Functional neuroimaging of odor imagery. NeuroImage. 2005;**24**:791-801

[36] Bensafi M, Sobel N, Khan RM. Hedonic-specific activity in piriform cortex during odor imagery mimics that during odor perception. Journal of Neurophysiology. 2007;**98**:3254-3262

[37] Rinck F, Rouby C, Bensafi M. Which format for odor images? Chemical Senses. 2009;**34**:11-13

[38] Arshamian A, Larsson M. Same same but different: The case of olfactory imagery. Frontiers in Psychology. 2014;**5**:1-8

[39] Frasnelli J, Lundström JN, Boyle JA, Djordjevic J, Zatorre RJ, Jones-Gotman M. Neuroanatomical correlates of olfactory performance. Experimental Brain Research. 2010;**201**:1-11

[40] Yao L, Pinto JM, Yi X, Li L, Peng P, Wei Y. Gray matter volume reduction of olfactory cortices in patients with idiopathic olfactory loss. Chemical Senses. 2014;**39**:755-760

[41] Kobayashi M, Sasabe T, Shigihara Y, Tanaka M, Watanabe Y. Gustatory imagery reveals functional connectivity from the prefrontal to insular cortices traced with magnetoencephalography. PloS One. 2011;**6**:e21736

[42] Rolls ET. Taste and smell processing in the brain. Handbook of Clinical Neurology. 2019;**164**:97-118

[43] Nakata H, Sakamoto K, Kakigi R. Meditation reduces pain-related neural activity in the anterior cingulate cortex, insula, secondary somatosensory cortex, and thalamus. Frontiers in Psychology. 2014;**5**:1489

[44] Zeidan F, Baumgartner JN, Coghill RC. The neural mechanisms of mindfulness-based pain relief: A functional magnetic resonance imaging-based review and primer. Pain Reports. 2019;**4**:e759

[45] Gibson J. Mindfulness, interoception, and the body: A contemporary perspective. Frontiers in Psychology. 2019;**10**:2012

[46] Tang YY, Lu Q, Feng H, Tang R, Posner MI. Short-term meditation increases blood flow in anterior cingulate cortex and insula. Frontiers in Psychology. 2015;**6**:212

[47] Hölzel BK, Carmody J, Vangel M, Congleton C, Yerramsetti SM, Gard T, et al. Mindfulness practice leads to increases in regional brain gray matter density. Psychiatry Research. 2011;**191**:36-43

[48] deCharms RC, Maeda F, Glover GH, Ludlow D, Pauly JM, Soneji D, et al. Control over brain activation and pain learned by using real-time functional MRI. Proceedings of the National Academy of Sciences of the United States of America. 2005;**102**:18626-18631

[49] Krause F, Kogias N, Krentz M, Lührs M, Goebel R, Hermans EJ.

Self-regulation of stress-related large-scale brain network balance using real-time fMRI neurofeedback. NeuroImage. 2021;**243**:118527

[50] Sitaram R, Cara A, Veit R, Gaber T, Ruiz S, Birbaumer N. Volitional control of the anterior insula in criminal psychopaths using real-time fMRI neurofeedback: A pilot study. Frontiers in Behavioral Neuroscience. 2014;**8**:344

[51] Craig AD. How do you feel--now? The anterior insula and human awareness. Nature Reviews. Neuroscience. 2009;**10**:59-70

[52] Damasio AR. The somatic marker hypothesis and the possible functions of the prefrontal cortex. Philosophical Transactions of the Royal Society of London. Series B, Biological Sciences. 1996;**351**:1413-1420

[53] Caria A, Sitaram R, Veit R, Begliomini C, Birbaumer N. Volitional control of anterior insula activity modulates the response to aversive stimuli. A real-time functional magnetic resonance imaging study. Biological Psychiatry. 2010;**68**:425-432

[54] Yao S, Becker B, Geng Y, Zhao Z, Xu X, Zhao W, et al. Voluntary control of anterior insula and its functional connections is feedback-independent and increases pain empathy. NeuroImage. 2016;**130**:230-240

[55] Kabat-Zinn J. Mindfulness Based Stress Reduction (MBSR). Constructivism in the Human Sciences. 2003;**8**:73-83

[56] Li W, Luxenberg E, Parrish T, Gottfried JA. Learning to smell the roses: Experience-dependent neural plasticity in human piriform and orbitofrontal cortices. Neuron. 2006;**52**:1097-1108

[57] Kollndorfer K, Fischmeister FP, Kowalczyk K, Hoche E, Mueller CA, Trattnig S, et al. Olfactory training induces changes in regional functional connectivity in patients with long-term smell loss. Neuroimage Clinic. 2015;**9**:401-410

[58] Molnar-Szakacs I, Uddin LQ. Anterior insula as a gatekeeper of executive control. Neuroscience and Biobehavioral Reviews. 2022;**139**:104736